T5-ANA-522

Collector's Guide to

Herkimer Diamonds

Michael R. Walter

4880 Lower Valley Road • Atglen, PA 19310

Schiffer Earth Science Monographs Volume 18

Other Schiffer Books on Related Subjects:

Collector's Guide to Quartz and Other Silica Minerals.
Robert J Lauf
978-0-7643-4161-8. $19.99

Collector's Guide to the Zeolite Group.
Robert J. Lauf
978-0-7643-4675-0. $19.99

The Collector's Guide to Silicate Crystal Structures.
Robert J. Lauf
978-0-7643-3579-2. $19.99

Copyright © 2014 by Michael R. Walter

Library of Congress Control Number: 2014947473

All rights reserved. No part of this work may be reproduced or used in any form or by any means—graphic, electronic, or mechanical, including photocopying or information storage and retrieval systems—without written permission from the publisher.

The scanning, uploading, and distribution of this book or any part thereof via the Internet or via any other means without the permission of the publisher is illegal and punishable by law. Please purchase only authorized editions and do not participate in or encourage the electronic piracy of copyrighted materials.
"Schiffer," "Schiffer Publishing, Ltd. & Design," and the "Design of pen and inkwell" are registered trademarks of Schiffer Publishing, Ltd.

Designed by Molly Shields
Type set in Bickham Script Pro/Cambria

ISBN: 978-0-7643-4710-8
Printed in China

Published by Schiffer Publishing, Ltd.
4880 Lower Valley Road
Atglen, PA 19310
Phone: (610) 593-1777; Fax: (610) 593-2002
E-mail: Info@schifferbooks.com

For our complete selection of fine books on this and related subjects, please visit our website at www.schifferbooks.com. You may also write for a free catalog.

This book may be purchased from the publisher. Please try your bookstore first.

We are always looking for people to write books on new and related subjects. If you have an idea for a book, please contact us at proposals@schifferbooks.com.

Schiffer Publishing's titles are available at special discounts for bulk purchases for sales promotions or premiums. Special editions, including personalized covers, corporate imprints, and excerpts can be created in large quantities for special needs. For more information, contact the publisher.

Dedication

This book is dedicated to my father, Jay Walter, who has always loved collecting Herkimer diamonds. At age 75, he is still swinging a 20-pound sledgehammer (see Fig. 83).

Contents

Foreword from the Series Editor

Herkimer diamonds are instantly familiar to practically every mineral collector. Not only are they easily recognizable for their simple beauty, variety, and perfection, but they are plentiful and located in areas where the amateur can actually collect them!

In this volume, Michael Walter provides an in-depth look at these fascinating objects and offers the kind of essential collecting guidance that only a very seasoned field collector can give. The author has years of experience collecting in New York State and shares his knowledge, along with stunning photos of crystals and detailed locality data. Collectors who want to dig for themselves will find this book to be an invaluable guide for where to go, how to work the rock for maximum success, and how to protect, preserve, and document the specimens they have found. Collectors who must be content to buy specimens or who can't travel to New York will also benefit as the book will help them to better understand and appreciate the specimens they have in their collections.

Robert J. Lauf
Series Editor

Preface

With the replacement of a single word, *minerals*, Edward S. Dana's preface from his book titled, *Minerals and How to Study Them*, is the perfect lead up for this book:

"The author has occupied some hours, which could not be devoted to more serious labor, in preparing this little book, in the hope that it might serve to encourage those who desire to learn more about *minerals*, and also to increase the number of those whose tastes may lead them in this direction." (Dana, 1895; italics added)

Substitute *Herkimer diamonds* for *minerals,* and it sets the stage nicely.

Today, writing about a topic that one is passionate about is not so much a chore as it is a challenge. The inability to field-collect minerals during long New York State winters has allowed me time to take on this challenge.

There is great enjoyment in bringing a critical mass of information together in one place and providing it to those who have similar interests. Writing this book about collecting Herkimer diamonds, what could be viewed as a narrow topic by some, provided numerous challenges. Actually, I found it to be a very large topic and had to limit the range of included content. As a result, I wrote the history section not to be comprehensive, but only introductory. The locations I present exclude several whose owners did not want them to appear in print. The many things I may not know about techniques and tools are obviously absent. I found other hurdles as well. How do I sort historical fact from fiction and myth? Is expert opinion freely available? Tools and techniques for collecting Herkimer diamonds vary, so which are best? All these things presented themselves as challenges; however, I hope my passion for this topic comes through loud and clear. I have tried to limit personal opinion and speculation, but not scientific fact.

I have relied on more than twenty-five years of experience at field-collecting Herkimer diamonds and my general success at finding them is the core knowledge of this book. By no means is this a claim of superior professional expertise, unrivaled ability, or comprehensive knowledge of the topic. There are many other individuals with longer histories and greater levels of success at finding Herkimer diamonds. Some even have similar interests in writing about collecting Herkimer diamonds or recording their personal experiences in finding them. I decided to make my contribution to our understanding by writing this book.

The task of finding Herkimer diamonds in the field has become increasingly difficult with the passage of time. Most of the easy-to-reach specimens have been gathered and now, more than ever, information on how to collect Herkimer diamonds effectively is both wanted and needed. Experienced Herkimer diamond collectors can spend a large percentage of their collecting day fielding questions about tools, techniques, where to find Herkimer diamonds, and successes from the more casual visitors who want to find their own crystals.

I hope this book will enable others to have success at casual collecting, or mining, their own Herkimer diamonds. Penetrating the Little Falls Dolostone to find Herkimer diamonds is always going to be challenging. By looking at appropriate locations with an educated eye and by using the tried and true tools and methods presented here, I hope readers will have their own good experiences.

Acknowledgments

I am grateful to the Canadian Museum of Nature, Steve Chamberlain, Jeff Fast, Jay and John Medici, M. and A. Mizutani/ Nakanishi, the New York State Museum, Stuart Strife, Scott Wallace, and Jay Walter for providing access to their specimens for photography and for providing photographs themselves. I thank Bob Morgan for helpful discussions about twinning. I thank Mark-Paul Serafin for assistance with designing and implementing the graphics. I thank Helen and Steve Chamberlain for their editing and encouragement.

Unless otherwise noted, all specimens in this book are from the author's personal collection, and many he personally collected. All photographs were taken by the author, except as noted.

Chapter 1: Introduction

Herkimer diamonds, a stunning form of quartz crystals, are probably more familiar to the general population than any other mineral from New York State. Moreover, they are particularly prized by mineral collectors, figure in the region's history, and are well represented in the scientific literature. Similar quartz crystals from other sites around the world are usually portrayed as being like Herkimer diamonds and their fame has even created a small-scale tourist industry in parts of the state where they occur. Why is this?

Herkimer diamonds are actually the mineral quartz, silicon dioxide (SiO_2), while diamond is pure carbon (C). The name Herkimer diamond reflects their being discovered by European settlers in what became Herkimer County. The term "diamond" is a misnomer that resulted from their having a somewhat unusual habit for quartz crystals in being short and doubly-terminated and their having a colorless, flawless perfection in small crystals, much like the clarity of a fine faceted diamond. Traditionally, quartz forms long, slender, prismatic crystals. The differences between short and stubby and long and slender are quite obvious, so Herkimer diamonds got a special varietal name. To the general populace, Herkimer diamonds are easily recognized, and they are sometimes even confused with faceted diamonds. Small, exceptionally clear Herkimer diamonds superficially look so much like real diamonds that they are often used in jewelry, usually in their natural crystal form.

Because Herkimer diamonds are quartz crystals, they have the physical and chemical properties of quartz. Like all other forms of quartz, their hardness is placed at 7 out of 10 on the Mohs hardness scale, with actual diamond holding the top spot at 10 as the hardest substance known on earth. Numerous sources have perpetuated the mistaken idea that Herkimer diamonds have a hardness of 7.5 beyond that of normal quartz. This is simply untrue. Perhaps this idea comes from the fact that the hardness of quartz can vary depending on which crystal face you scratch, the amount of imperfections found within the crystalline structure, or that a quartz specimen can scratch another quartz specimen. Whatever the origin of this myth, the exceptional crystals from this region do not possess exceptional hardness when compared to quartz from elsewhere around the world.

Their specific gravity is moderate at 2.65. They are glassy in appearance and have a conchoidal fracture much like glass. The crystals sometimes show very light striations, or parallel grooves, on their surfaces. Herkimer diamonds are most noted for being transparent and colorless. Sometimes a smoky tint, grading all the way to black, is seen due to finely disseminated inclusions of hydrocarbon. This hydrocarbon is often referred to as anthraxolite, which is a geological term, not a formal mineral or rock name. Indeed, anthraxolite is the most common term used for the hydrocarbon substance that forms in and around Herkimer diamond crystals in the region. Most crystals show the classic 18-sided, doubly-terminated, hexagonal form. They range in size from that of fine sand particles to the size of a football. The largest, perfectly clear Herkimer diamond is reputed to be in the collection of a New Jersey family and is said to measure in

Fig. 1. *Examples of quartz. Left: A typical prismatic quartz crystal from Ron Colman's mine, Hot Springs, Arkansas. 11 cm. Right: Herkimer diamond. Herkimer Diamond Mines Resort, Middleville, NY. 2.5 cm.*

at over 10 centimeters. Examples of perfectly clear crystals seldom exceed 4 centimeters, however.

Quartz is one of the earth's most common rock-forming minerals. Most children will recognize the mineral and can identify it on sight almost as easily as they can pyrite "fool's gold." Although a common mineral, quartz crystals with as perfect form and clarity as Herkimer diamonds are rare.

Virtually every museum mineral collection contains fine examples of Herkimer diamonds. Any mineral collector, from the beginner to the advanced connoisseur, will be familiar with them and own examples. Indeed, the general population has been collecting these crystals as simple keepsakes for centuries. It is appropriate that the New York State legislature is in the process of designating the Herkimer diamond form of quartz as the New York State mineral to join the garnet as the New York State gemstone.

Many individuals advance beyond buying or trading for these desirable crystals and actually go to the field to collect their own specimens. As a result, there are a number of commercial operations in New York that cater to collectors and tourists. During the summer, collectors from all over the world flock to the Mohawk Valley region in Central New York to try their luck at finding Herkimer diamonds.

The fact that these beautiful objects form in solid rock, often only centimeters from our reach, brings out the adventurous nature of many people. That treasure-hunting mentality is often what draws so many from so far to try their luck. The allure of the search is compelling enough for many to go collecting time and time again. Collecting Herkimer diamonds has an addictive quality for some that has no equal. Massive amounts of time and resources may be devoted to the pursuit of these crystals.

Herkimer diamond crystals actually show significant diversity in form and appearance from location to location. Tiny crystals of amazing clarity are the most popular with collectors, but large crystals showing a greater degree of internal fracturing, crystals that form directly on the surrounding rock, or oddities involving materials found within the crystals are not uncommon. It is this diversity

Fig. 2. Left: *Herkimer diamond cluster. Ace of Diamonds mine, Middleville, NY. Collected 1990. 5 cm.* Right: *Faceted Herkimer diamond weighing 5.5 carats.*

Fig. 3. *Examples of Herkimer diamonds.* Top: *Cluster composed of two Herkimer diamonds. Treasure Mt. mine, Little Falls, NY. Collected 2002. 3.2 cm.* Bottom: *150 small Herkimer diamonds. Herkimer Diamond Mine Resort, Middleville, NY. Jay Walter collection.*

that makes Herkimer diamonds so fascinating to so many. Objects created by nature that are hundreds of millions of years old, but so perfect that they appear to be cut and polished, draw our curiosity. Enhancing this interest are the stories of the discoveries that other collectors have made in the past. Both stories in print and those spread by word of mouth help build the legends that go with important Herkimer diamond specimens. There are few things as motivational as actually watching someone open a pocket containing hundreds of gemmy Herkimer diamonds and sharing their joy of discovery. Wiping the cool mud from the surfaces of a baseball-sized crystal creates an indelible memory that will make the heart race and convert casual interest into passion. Acting on that passion and creating an appreciation for the diversity found in these crystals is the purpose of this book.

For most, learning to collect these fine crystals is not something passed from parent to child (although there are some well-documented exceptions) or discovered in an academic setting. The proper knowledge needs to come from those who have expertise based on first-hand experiences. Although there is no substitute for actual field experience,

reading about the methods behind the collecting of Herkimer diamonds can provide the beginner with some very useful background on the topic. More in-depth readings can even go as far as providing details that will take the experienced collector to the next level of skill.

Even though an entire body of information has developed around the successful collecting of these crystals, very little has been published that helps the novice become expert. There have been articles describing specific localities and others on the attributes of the crystals; however, no comprehensive treatment describing, in detail, what needs to be done to have success at field-collecting Herkimer diamonds has yet been published. Likewise, nothing fully describing the great diversity of crystal forms, varieties, and associated minerals has ever been published. This book aims to provide the background and methods that should assist anyone wishing to find Herkimer diamonds on their own no matter what their level of expertise. Information is presented to a level of detail that goes beyond that of any previous publication and is accompanied by illustrations that should give the understanding needed to assist anyone in becoming a successful collector of Herkimer diamonds.

Fig. 4. *The Greens, fifty-plus years of field collecting and still smiling, pictured here at their long-term claim. Diamond Acres mine, Mohawk Township, NY. September 4, 2011.*

Chapter 2:
History of Herkimer Diamonds

Native American history shows an intimate connection between the inhabitants of the Mohawk Valley and the Herkimer diamonds found in that region. Although not a proper Native American name, Mohawk is derived from what the Mohican Indians living in the upper Hudson Valley called their enemies to the west, *Mohowaug*, "they eat living creatures." However, the natives referred to themselves as *Kanyenkehaka* and the valley they lived in as *Kanyenka*.

Kanyenka translates to "the place of flint"; however, this meaning has been brought into question by several scholars. Most prominent in this discussion is Dean R. Snow, Professor and Head of Anthropology at Pennsylvania State University. He writes:

> "They were known to themselves and to the other Iroquois nations as the Kanyenkehaka, the people of Kanyenka (also spelled Ganienkeh). This has been translated "the place of the flint," but the flint (or more properly chert) sources in Mohawk country were not particularly sought after. More important were the clear quartz crystals now called Herkimer diamonds, which could be quarried in a few local mines and abound on Mohawk village sites. These were highly valued by Iroquois and other nations. Kanyenka was more likely "Place of the Crystals." (Snow, 1996)

As there are no known deposits of flint in the Mohawk Valley, the correct translation is more likely "place of the crystals." It is quite obvious to anyone knowing the geology, mineralogy, and native history of the area that these people valued, sought out, traded, and were known for Herkimer diamonds and not flint.

Herkimer diamonds are known to have been symbolically important as amulets and were sometimes used to make tools. The Mohawks supplied these crystals to other natives up until 1614, when they were replaced by Dutch-made glass beads.

With European occupation, the Herkimer name was brought to the region by the Johan Jost Herkimer family, who established a home and trading post on the south side of the Mohawk River. The small community grew and, during the French and Indian War, was fortified in order to defend its new residents. The name Herkimer became accepted as a name for the community. Later, after the Revolutionary War, the county of Herkimer was formed in 1791, and it is from one of these two sources, the family name or the county name, that the term Herkimer diamond arose.

Herkimer diamond is a generic name given to the doubly-terminated quartz crystals, which were first noticed by European settlers in exposed outcrops of dolostone in the Little Falls area along the Mohawk River Valley. Other names only add to the confusion. Today commercial operations within Herkimer County often want to differentiate their crystals with special names, e.g. Middleville diamonds and Little Falls diamonds. None of these names appears to be a registered trademark; however, some treat the names as proprietary. Herkimer diamonds found outside of Herkimer County are often considered by purists not to be "Herkimers" at all.

Although there is little agreement on the strict definition of what qualifies as a Herkimer diamond, an inclusive approach is taken here to consider doubly-terminated quartz crystals of similar origin in geologically related strata, including the Beekmantown dolostone and the dolostones of the Theresa group. Thus, deposits in New York and southern Ontario that are outside the historical geographical region for Herkimer diamonds are included. However, the central focus will remain on occurrences in the more constricted region of Herkimer, Fulton, and Montgomery Counties.

There are many other world-wide locations where doubly-terminated quartz crystals with similar characteristics occur and use the term "diamond" in their names. International locations producing quartz under the term "diamond" include: Bohemian diamonds, Austria; Carrara diamonds, Italy; Diamantes de San Isidro, Spain; Marmarosch diamonds, Hungary and Romania; Mirabeau diamonds, France; and Schaumburg and Stolberg diamonds, Germany. Here in the United States there are other locations also using the same terminology: Cape May diamonds, New Jersey; Lake County diamonds, California; Pecos diamonds, New Mexico; and

Payson diamonds, Arizona. These will not be considered, as they are outside the region of interest both geographically and geologically. Of all these quartz crystals, Herkimer diamonds are the best known and highest in quality.

The modern mineralogical history of Herkimer diamonds is well documented. Dozens of books, journal articles, video clips, and news reports of variable accuracy are available. The references listed at the end of this book will give the reader an extensive base to draw from, but for now, only one of the

Fig. 5. *Chinese "Herkimers" to 3.5 cm.*

Fig. 6. *Swiss "Herkimer". 2 cm.*

older and one of the more contemporary publications will be mentioned in greater detail.

An older, detailed, investigative report regarding Herkimer diamonds comes from the *Mineralogy of New-York*, by Lewis C. Beck. Beck was a professor of chemistry and natural history at Rutgers College in New Jersey and was commissioned by the Governor of New York, William L. Marcy, to produce a comprehensive listing and description of the mineralogical wealth of New York State. In 1842, the 536-page document was published and included the first scientific discussion of Herkimer diamonds. The descriptions of minerals in this work are broken down by species and then by counties.

On pages 261-264, Herkimer diamonds are discussed under the mineral quartz. The first mention of these unusual quartz crystals falls under Herkimer County. Beck's crystallographic drawings were extensive, including figure numbers 130 through 161 (see Appendix C). The following excerpt may be, in part, where the term Herkimer diamond found its literary coinage:

> "Uncommonly beautiful specimens of rock crystal, perfectly transparent, have been found in various parts of this (Herkimer) county, especially at Middleville, Fairfield, Little-Falls, Salisbury and Newport. They are found sometimes in cavities in the calciferous sandstone; and at others, they lie loosely in the sand, probably produced by the disintegration of the rock. Mr. Vanuxem has remarked that there is probably no locality in the world which produces more perfect or more beautiful quartz crystals than Middleville; and with an equal refractive power they would rival diamond." (Beck, 1842)

Beck credits Professor J. Hadley of the Geneva Medical College as being the first to bring these quartz crystals to public attention in an article written for the *New-York Medical and Physical Journal* (II, 132). Beck's description of these fine crystals is scientifically detailed, including locations that were producing the specimens at that time; accurate crystallographic renderings, including rare forms; and physical details of the mineral, including general information like crystal size, and more detailed findings, such as inclusions and associated mineral species.

A recent summary account of Herkimer diamonds can be found in the 2008 book, *American Mineral Treasures*, published by Lithographie, LLC of East Hampton, Connecticut, that includes detailed descriptions of 44 of the most important mineral occurrences in the United States. An interesting and informative overview of Herkimer diamonds was written by the father-son team of John and Jay Medici of Ohio. The Medicis are well-known field collectors, not only of Herkimer diamonds, but also of many other minerals from numerous places in the United States and Canada. In preparing their chapter, they consulted with many other Herkimer diamond miners, mine owners, museum curators, and mineral collectors and, thereby, created a summary of Herkimer diamonds and the personalities involved in collecting them. In their two-paragraph historical overview, they note both the pre-historical usage of Herkimer diamonds by Native Americans to make arrowheads and the first major encounter of Herkimer diamonds by non-native settlers during the construction of a lock on a private canal at Little Falls in 1792.

These two sources are mentioned to highlight the fact that when the mineralogy of New York State is scrutinized it becomes clear that many believe the Herkimer diamond to be of great mineralogical importance. New York State is dotted with famous mines and occurrences that have produced some of the most spectacular mineral specimens in the world, both historical and contemporary. When those most passionate about the subject of specimen mineralogy focused on the issue of choosing a mineral to represent New York State for *American Mineral Treasures*, the Herkimer diamond was selected above all others.

Depositional Environment

The Little Falls dolostone formed from sediments deposited in the late Cambrian along the southern margin of what are now the Adirondack Mountains. This continental margin setting was along the edge of an inland sea that covered much of present day North America, but not the Canadian Shield, of which the Adirondack Mountains are an extension. The subtidal and intertidal environments (Fisher, 1977, Zenger, 1981) have been deduced from the comparison with modern day environments. However, Zenger notes "well over 100 lithologies," so any statements regarding specific depositional conditions in the literature are only summaries.

In general, there was a combination of carbonate and silica (sands and chert) deposition in a dynamic environment. A modern day environment of comparable similarity might be the Persian Gulf (Schneider, 1975, p. 213). A specific model explaining all the complexities of the sedimentology of Little Falls is lacking and may never be fully developed or understood. Whatever the paleoenvironment, it is known that silica in the form of sands and cherts provided ample material to create the prolific deposits of quartz crystals — Herkimer diamonds — found in various layers of the Little Falls.

Fossils are uncommon in these deposits. This is understandable given the lack of complexity, including hard parts, and diversity of species present at this early time in Earth's history and the processes of dolomitization and solidification of the limestone that were not conducive to fossil preservation. As a result, setting specific ages and environmental conditions using index fossils or the matching of species to appropriate biomes is difficult. However, cyanobacteria, commonly known as blue-green algae, are well represented in some layers and formed accretionary structures known as stromatolites. These fossil colonies are similar to modern colonies like those found in Shark Bay, Australia and elsewhere. It is within these horizons that the pockets in which Herkimer diamonds sometimes formed are found.

History of the Geologic Study of the Dolostone

Because of the complex lithologies and the scarcity of index fossils in the Little Falls Dolostone, geologists only slowly reached a consensus on how to group these layers in the geological column and how to assign their geological age. In 1905, H. P. Cushing mapped the Beekmantown formation and Little Falls Dolostone in the Little Falls quadrangle and assigned them to the Lower Silurian (Ordovician) geological time period. In 1910, E. O. Ulrich and Cushing restricted the Little Falls Dolostone of the Mohawk Valley (type region) to the lower part of the Beekmantown group and assigned it to the "late Cambrian" age. At present, the consensus view of the U. S. Geological Survey is that the Beekmantown group is represented by the Little Falls Dolostone and the Hoyt Formation in the Mohawk Valley. The age of the various layers of the Little Falls Dolostone containing Herkimer diamonds is taken as late Cambrian (about 495 million years ago), but the age of the quartz crystals themselves is inferred to be Carboniferous, as summarized by H. S. Muskatt and V. P. Tollerton, Jr., in 1992. However, not all occurrences of Herkimer diamonds sit in the Little Falls Dolostone; some are hosted in other divisions of the Beekmantown and Theresa Groups.

Petrology

The thickness of the Little Falls Dolostone in the Mohawk Valley has been mapped and shown to vary between 16 and 75 meters depending on its location. Regionally, it tapers in a northeasterly direction as it approaches the Adirondack Mountain region. The Little Falls varies in composition throughout, but is dominated by dolostone that is dark gray in color, finely to coarsely crystallized, and contains varying amounts of silica. This description, however, does not paint a complete picture. There are also numerous horizons rich in sand, sandstones, chert, and conglomerates that could have served as silica sources for the formation of Herkimer diamonds. Some horizons are vuggy while others are dense. The dolostone sits unconformably on the Precambrian rock of the Adirondack Mountains, mostly syenite in the Herkimer region, and is highly jointed. Later, regional faulting occurred and was extensive in some areas. The complexities of the region were well described and mapped by D. H. Zenger in 1981.

Age

Herkimer diamond crystals, like all other crystals found in cavities in rock, are younger than the rock that contains them. The question remains as to how old they are. Most current literature operates under two well-documented premises: first, that they formed slowly over a long period of time by crystallizing from unsaturated solutions indicated by their exceptional clarity and large size; and second, that they are younger than late Cambrian age.

Two different studies of temperature have essentially settled the age of Herkimer diamonds. Dr. Edwin Roedder, known for his extensive study of fluid inclusions in various crystals, studied Herkimer diamonds and determined the composition of the fluids from which they formed and their temperature of formation. Dr. Gerald M. Friedman studied the depths of burial of the northern Appalachian Basin during the Paleozoic period. He determined that the highest temperature, correlating with the maximum depth of burial, occurred in the Carboniferous period. His highest temperatures for the Little Falls dolostone (around 200°C) match pretty well with Roedder's temperatures of formation for Herkimer diamonds. Taken together, these two lines of investigation strongly suggest that Herkimer diamonds are about 175 million years younger than the rock in which they occur, 325 million-year-old crystals in 500 million-year-old rock.

Because the enclosing Little Falls Dolostone is about 500 million years old, a number of subsequent geological events have been implicated in the formation of Herkimer diamonds. Three periods of mountain building (orogenies) to the east all could have pushed solutions to the west. The Taconic orogeny, which ended around 440 million years ago, likely preceded the formation of the Little Falls; however, both the Acadian orogeny (400 to 325 million years ago) and the Alleghenian orogeny (260 to 325 million years ago) potentially could have been the source of heated mineralizing solutions. Such a mechanism is widely believed to be responsible for the mineralization found in cavities in the younger Lockport Dolostone that outcrops to the west of the exposures of the Little Falls Dolostone. However, the mineralization in the Lockport Dolostone is a relatively different set of minerals. Nothing remotely like Herkimer diamonds occurs in the cavities, and there is no consistent association of clear quartz crystals and hydrocarbons.

The organic materials required to create the hydrocarbons now found in the pockets and as inclusions in Herkimer diamond crystals may have been trapped during the initial burial of the sediments that formed the dolostone or may have migrated into the voids from other sources or both. A unique, large vein of hydrocarbon was found by Ronald Backus in November of 1967 (Labuz, A. L., 1969) on Route 167, about two miles southwest of Little Falls, "...a horizontal vein about a foot wide, 15 feet long, and undetermined depth, trending north 25 degrees east, terminating to two branches of 1 and 2 ft." Massive hydrocarbon to several inches and unique spheroids were recovered. The vein was approximately four to five feet below the road surface. This important find points to the fact that, at one time, the hydrocarbons in this rock were concentrated and somewhat mobile. It is easily seen on a small scale in and around vugs; however, large scale examples such as this are very rare.

Paragenesis

The order of crystallization of the various minerals found in cavities containing Herkimer diamonds might seem to vary from locality to locality, mostly due to many cavities having an incomplete sequence of deposited minerals. However, some general trends emerge when various occurrences are compared. Hydrocarbon in the form of what is referred to as anthraxolite is generally first in the paragenesis. Many Herkimer diamonds contain inclusions of black anthraxolite. Clouds of tiny hydrocarbon particles account for the smoky brown color of some Herkimer diamonds. At a few localities, such as the St. Johnsville quarry, quartz, dolomite, and calcite all contain plentiful particles of hydrocarbon. Dolomite seems to be next in the order of deposition, often lining the walls of the cavities. It is often obvious that quartz followed dolomite in the order of formation, but usually preceded calcite, which is often last. Sulfides seem to occupy several positions in the paragenesis. Sometimes, for example, pyrite crystals are found as inclusions in quartz, but at other times, sulfides are found on the surfaces of calcite crystals.

How Herkimer Diamonds Formed

The mechanism for the formation of Herkimer diamonds has been hard to determine. At least two things have contributed to this gap in our knowledge. Sedimentary petrologists tend to concentrate on the nature, age, and formation of the rock itself since the minerals in cavities usually formed later. In addition, the mechanism most likely to explain the formation of Herkimer diamonds and their consistent association with hydrocarbons was only discovered and published in 1987. Is it any wonder, therefore, that so much confusion, so many different theories, and such fervor in defending one's favorite theory have characterized discussions among Herkimer diamond miners, mineral collectors, and scientists? Some of these theories do not deserve further mention. Others are simply observational and are untested using any acceptable scientific model. Several are described in previous books on this topic or written about on active web sites. These models will not be discussed here, and it is recommended that this type of conjecture be avoided if one wishes to better understand Herkimer diamond formation.

One of the more recent areas of study has been the involvement of organic mechanisms in the formation of Herkimer diamonds. The general trend in earth science indicates that there is far more overlap of the biosphere and the lithosphere than was previously understood. The idea that Herkimer diamonds, or any "proper" minerals, could have formed by mechanisms involving organic molecules does not fit well with how a mineral has been defined by American mineralogists for more than 200 years (i.e., having formed due to inorganic processes). The Manual of Mineralogy in its 19th edition, 1977, modified the previous definition to state that they are *usually* formed by inorganic processes. These mechanisms have been more widely studied and accepted in recent years and are providing important evidence regarding crystal formation, generally, and the origins of Herkimer diamonds, specifically.

How the interaction between organic molecules and quartz was discovered is interesting. Dr. Donald Siegel and his doctoral student, Phillip Bennett, were part of a larger effort to study what happens when an oil spill occurs on land. In particular, a heated oil pipeline near Bemidji, Minnesota, burst and spilled large quantities of crude oil onto glacial sand in a wilderness area. The site was set aside for scientific study. Siegel and Bennett were looking at the concentration and size of organic molecules in the spill at various distances from the leak. They found that the hydrocarbons in the crude oil were being broken down by bacteria and becoming shorter and shorter molecules over time. The acidity (pH) of the spilled oil was neutral. Near the spill, there was little dissolved silica (as would be expected because quartz sand is not soluble in either water or crude oil at normal atmospheric pressure and temperatures); however, farther from the spill where the crude oil was partially degraded, there was a very large amount of dissolved silica. Still further from the spill where the crude oil was even more degraded, the amount of dissolved silica again decreased to low levels. When they examined the grains of sand, they found that the grains were being dissolved where the dissolved silica was rising and that new quartz was depositing on the grain surfaces where the dissolved silica was

falling. Their observations and subsequent studies led to the realization that humic acids, organic molecules commonly present around the roots of plants in soil, can dissolve silica and hold it in solution at surface temperatures and neutral acidity. This breakthrough was published in the high-profile scientific journal *Nature*, to be followed eight months later by a discovery published in *Geology* by Friedman showing that the region's rock may have been buried at greater depth than previously noted. His study shows burial depths to 7 km existed and would have lead to temperatures at maximum burial upward of 200 degrees Celsius. These two new sets of facts were synthesized into a new theory for the formation of Herkimer diamonds by Drs. S. C. Chamberlain and R. V. Dietrich (Chamberlain, 1988; Dietrich and Chamberlain, 1989).

A contemporary summary of the origin of Herkimer diamonds begins at the bottom of a shallow sea about 495 million years ago. The material that would later form the Herkimer diamonds was probably deposited as a waxy organic material along with quartz sand and masses of pyrite formed by bacterial sulfate reduction. All of this was encased in a rock that is made up of two carbonate minerals, dolomite and calcite. As the Little Falls Dolostone was slowly buried by new sediments, its temperature slowly rose from perhaps 20°C to 175°C or slightly higher. This burial took about 200 million years and reached a maximum depth of 5 to 7 kilometers about 300 million years ago. As the temperature slowly rose, molecules from the organic material that had been holding quartz dissolved from the sand in solution were broken apart by thermal splitting. This caused the quartz to come out of solution very slowly and resulted in very slow growth of quartz crystals of exceptional clarity—Herkimer diamonds. All that remained, thereafter, was for the overlying sediments to weather away, exposing the Little Falls Dolostone at the surface.

It is likely that additional details will be added to this mechanism of formation as research continues; however, that the broad outlines are correct is reinforced by the realization that there are many examples of very clear quartz crystals with similar fluid inclusions and association with hydrocarbons in various sedimentary rocks worldwide. Of these, however, Herkimer diamonds are surely the best known and form the largest deposit.

Chapter 3:
Why Collect Herkimer Diamonds?

Why collect minerals? In early times, the wealthy, with the luxury of free time and expendable funds, were able to collect minerals, along with other natural history objects, and amass collections of significance. To them, in most cases, the importance of their collections reflected their interest in the science of our natural world. Minerals were often only one small part of that overall interest and minerals were only part of their natural history cabinet. For some, the interest centered only on minerals and important mineral collections were built.

In answering the more specific question "Why field-collect Herkimer diamonds yourself?", you may have the answer already. If you are reading this book, it is quite possible you have a level of interest that compels you to see the unseen or find that which has not yet been found. Many collectors describe the thrill of the hunt or the excitement associated with being the first human to lay eyes on that crystal as it comes from within the rock as motivations for collecting. Few mention the hard work, the injuries, the monotonous routine of breaking rock, the long hours, the long drives, the cost, and so on, as the reasons for their field-collecting Herkimer diamonds. It would seem that the positives are few while the negatives are extensive. Yet so many make the trip to these areas in central New York State from all reaches of the country and world. Some spend weeks, months, years, or entire lifetimes in the pursuit of these crystals. What is the allure?

Many are drawn to Herkimer diamonds by their great aesthetic appeal, whether they field-collect them, see them at a mineral show, observe them in a museum, or are exposed to them by a friend.

Thousands of specimens of Herkimer diamonds are available for sale in gift shops, at mineral shows, and in auctions and stores on the internet. They are so easy to obtain, why not just buy them? When all the costs associated with collecting them for yourself are considered, it becomes clear that it certainly might be less costly to just buy them.

The challenges of collecting them yourself are extensive. Travel to the rather remote region where they occur is necessary if you want to collect these crystals yourself. The rock is hard. This might sound obvious because all rock is more or less hard! The dolostone of the Little Falls formation goes beyond what most people think of as hard and needs a description of its own. The problem is there is no single word in our language that effectively conveys how difficult this rock is to break for a Herkimer diamond collector. The rock is literally

Fig. 7. Mark Hawkins, Terry Antwine, and Dennis Snyder collecting Herkimer diamonds. Crystal Grove mine. Fall 2009.

hard, tough, and not extremely brittle. The word "bulletproof" comes to mind and is often used by seasoned collectors. One might think of comparing the Little Falls Dolostone to *rock* as one might compare Kevlar to *cloth*. It is basically impossible to penetrate without learning special techniques.

The conditions under which one does the collecting are often brutal. Normally collecting occurs only during the favorable weather of summer, often in quarries exposed to relentless direct sunlight. If you've seen pictures or films of prison chain gangs breaking rock with sledge hammers, that is very close... just remove the leg shackles and it will be about right. If you are not in good physical shape, the work will seem overwhelming.

So why field-collect Herkimer diamonds yourself? Well, maybe it is as simple as your attraction to the thrill of the hunt or your being the first person to see that crystal in the soil or coming out of the rock. Little kids love to do it and they are seldom prepared for any of the challenges. Fortunately, there are tiers of difficulty in finding Herkimer diamonds. You can look for loose crystals among the broken rock or in the soil. Any beginner who can lift a four-pound crack hammer will find the crystals by breaking the smaller boulders. Only the most serious collectors mine the ledge to find new pockets. Over time, you can work your way from easier approaches to harder approaches. There are millions that have been found and millions more to be found.

This book may prove useful to you in your quest to collect your own Herkimer diamonds.

Fig. 8. *Jeremy Snell collecting Herkimer diamonds. Crystal Grove mine. Fall 2005.*

Chapter 4: The Mines

Dozens of past and present localities have produced the quartz crystals referred to as Herkimer diamonds from this large region of New York. Some of them are described in this chapter along with specimens collected from each. Directions to the localities featured here can be found in Appendices A and B. Their inclusion does not give the reader the right to enter any of these locations. Almost all sites where Herkimer diamonds can be found are tightly controlled by current property owners. Only some were open to the public when this book was written, and if they are open, fees are normally required for entry and the right to collect minerals. This warning should be taken seriously. Landowners have been known, even in modern times, to pull out the guns to protect their property rights.

Many current descriptions suggest that all Herkimer diamonds are basically alike, but such is not actually the case. The diversity of crystals found at the numerous collecting sites within the Herkimer diamond region is quite large. Specimens from a particular locality or small region often have particular characteristics. Experienced collectors who specialize in Herkimer diamonds are likely to be able to tell where an individual specimen was collected, even if it is from an undisclosed location, based on things such as specimen form, crystal habit, inclusions, nature of the matrix, and associated minerals.

Commercial Mines

Ace of Diamonds and **Herkimer Diamond Mines Resort** are two of the most frequented locations for collecting Herkimer diamonds. Previously, the Ace of Diamonds property was known as the Petrie or Tabor farm while the Herkimer Diamond Mines Resort land was known as the Herkimer Diamond Development

Corporation, Schrader's, or Van Atty's farm. The land just north of the Ace was owned by Claude H. Smith who wrote the locally popular book, *Let's Hunt for Herkimer Diamonds*. Due to their close proximity — literally side-by-side — within the village limits of Middleville, Herkimer County, New York, the two locations are being described under the same heading. Although under different management, both have extensive offerings and, with rare exception, specimens from the two locales are nearly indistinguishable. They are operated as fee locations and open to the general public. Visitors will be greeted by exposed walls of Little Falls Dolostone as high as five meters in an open quarry largely hand dug by collectors and miners attempting to reach large pockets containing both individual Herkimer diamonds and Herkimer diamond clusters. These pockets occur in layers parallel to the quarry floor and are referred to as dome or gooney pockets by

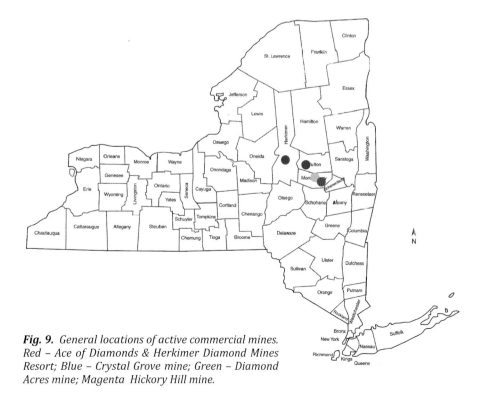

Fig. 9. *General locations of active commercial mines. Red – Ace of Diamonds & Herkimer Diamond Mines Resort; Blue – Crystal Grove mine; Green – Diamond Acres mine; Magenta Hickory Hill mine.*

experienced collectors. Many empty examples found by previous collectors will be evident at the intersection between the quarry's wall and floor. A sub-pocket layer is found below the larger pockets and is often referred to as ledge rock. Great quality crystals can be found in this densely-pocketed ledge rock, but they are usually smaller, although often clearer, than those found in the primary dome pockets. Some collectors who frequent these locations elect to break rock left behind by other collectors or dig in areas covered by dirt. Spots on the wall can be reserved and worked for multiple days, and there are extensive regulations regarding what tools may or may not be used to collect at these sites. Each location has an active website outlining most of the mine's amenities and programs.

Both Middleville locations are found within the Beekmantown Formation — more specifically, the Little Falls Dolostone. Here it is almost 70 meters thick in some places and it contains one of the most uniform layers of pockets to be found anywhere in the region. The primary pockets are usually clay filled and can approach two meters in diameter, although they are more commonly less than a meter in size. Within them are found individual crystals to 15 centimeters each and clusters commonly in the 10 to 20 centimeter size. More so than any of the other Herkimer diamond locations, the Ace and Herkimer Diamond Mines and Resort are known for their fantastic clusters. Clusters are almost always found as separate crystals that need to be reassembled if they are to be preserved as clusters, and most collectors are willing to undertake this painstaking task. Sceptered crystals are very rare here, but skeletal crystals are not uncommon and are usually large, to 15 centimeters, and often smoky. Over 1,000 individual crystals have sometimes been harvested from a single pocket but far fewer is the norm. Most crystals appear to form in or near the base of these primary pockets, where they are almost always found detached from the rock, lying along the pocket's base embedded in clay. Often a dome

of dolostone develops in the center of the base of the pocket and crystals form in a circular pattern around its center. As a result, some clusters are referred to as arches, partial encirclements, or as donuts — complete circles of crystals with an opening in the center. When successfully reassembled most dome pocket clusters are two-dimensional, with three-dimensional groupings being less common. In some pockets, larger colorless crystals are found on a drusy layer of small colorless or black quartz crystals.

Herkimer diamonds are occasionally encountered in rock above this primary pocket level. Although most collectors who work the walls of these locations simply strip off the upper layers and discard them, closer examination can yield good results. When crystals are encountered above the pocket level they are often well formed, very clear, and locked tightly in the dolostone.

Pockets from the ledge rock under the dome pocket layer sometimes contain nearly flawless Herkimer diamonds with calcite and dolomite as accessory minerals. One discovery produced nearly transparent nail-head form calcite in combination with very nice Herkimer diamonds. Dolomite crystals to a centimeter in length can form before or after the quartz. On occasion, pyrite rods are found in Herkimer diamonds from Middleville.

Crystals from Middleville normally show moderate internal cracking and often have inclusions of anthraxolite, especially in the base of the crystals where they form on or near the base of the pocket. Other inclusions include liquids and gas bubbles (less common) that sometimes move about within the crystal when it is tilted. Negative crystals are also present as inclusions in some crystals. Floaters are common and floater clusters, equally so. Smaller clear diamonds are exceptional and of such high quality that they are used in jewelry, often being referred to as jewelry stock. Black druse pockets are uncommon and provide what are considered

Fig. 10. *Getting close. South of Middleville, NY.*

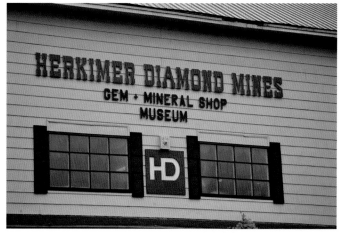

Fig. 11. *Herkimer Diamond Mines Resort facility. Middleville, NY.*

Fig. 12. *Hundreds of individual jewelry-grade, totally clear Herkimer diamonds. Ace of Diamonds mine, Middleville, NY. Crystals to 1 cm. Jay Walter collection.*

Fig. 14. *Freshly opened pocket. Large, clay-coated crystals are clearly visible. Ace of Diamonds mine, Middleville, NY. Field of view, approximately 60 cm. Jeffrey Fast photo.*

by most collectors to be the finest of all Herkimer diamond specimens — gemmy colorless Herkimer diamonds on a bed of black, anthraxolite-included quartz crystals.

These two popular sites draw in both inexperienced and true, hard-core miners from all over the United States and even other countries. The Middleville locations have been written about by clubs and mineralogical organizations worldwide, have been discussed in numerous books and journal articles, are described extensively on several active websites, and were even featured in a season-two episode of the Travel Channel's show, *Cash and Treasures*, hosted by Kristen Gum. They are major commercial business operations in the Middleville community.

These sites are easily found, with signs on the major highways. There are showers, food, camping, tool rental, gift shops, restaurants, and other necessities on-site or within a stone's throw of the digging areas. The Herkimer Diamond Mines Resort runs various outreach programs for youth: Scout jamborees for boys and girls as well as weekly summer camps are available to kids of various ages. Both locations are found on the west side of Route # 28, nine miles north of I-90 in the village of Middleville.

Fig. 13. *Large, tabular, smoky, skeletal crystal. Herkimer Diamond Mines Resort, Middleville, NY. 14.9 cm.*

Fig. 15. *Spectacular Herkimer diamond cluster. Herkimer Diamond Mines and Resort, Middleville, NY. 15 cm. Collected by Ken Silvy. Stuart Strife collection. Joe Budd photo.*

Crystal Grove is another commercial digging location. It is near Lassellsville, in the town of Oppenheim, in Fulton County, and like the Middleville locations, it has extensively excavated areas open to collecting. Crystals at this locality occur in the Galway (Theresa) Formation, which is a dolostone similar to the Little Falls Dolostone but younger in age. It grades into the Little Falls Dolostone towards the southwest. At Crystal Grove, there is also a large forested region where collectors can find Herkimer diamonds in the soil and in highly weathered bedrock. Unlike the two Middleville locations, there is no enormous quarry wall. Here the wall is much lower, usually two meters or less, but the rock is equally hard. Within a less definable layer, pockets can be found that are normally small, five to 15 centimeters in diameter. Sometimes the pockets have been invaded by clays, but equally common are very pristine pockets lacking coatings and stains. Seldom are the crystals larger than two centimeters in size. The crystals are most often found attached to the wall rock and are often attached to tiny crystals of dolomite or quartz. The druse coatings on the matrix rock add another attractive dimension to the specimens from this site. The druse here, like most other Herkimer diamond locations, is made of tiny clear quartz crystals seldom over a few millimeters in length that are often more prismatic, or elongated, than the typical Herkimer diamond. Matrix specimens feature Herkimer diamonds that

Fig. 16. *Large single crystal showing Dauphine habit (greatly enlarged rhombohedral face indicating a slower growth of that terminal face). Ace of Diamonds mine, Middleville, NY. 6.5 cm. Jay Walter collection.*

Fig. 17. *Freshly opened pocket. Herkimer Diamond Mines Resort, Middleville, NY. October 4, 2011. Total pocket yield: 1,600 Herkimer diamonds.*

Fig. 18. *View into the base of druse-lined pocket (Fig. 17) showing a mix of crystals, hydrocarbon, and druse. Herkimer Diamond Mine Resort, Middleville, NY. Field of view, approximately 12 cm.*

Fig. 19. *Portion of ledge showing randomly scattered pockets. Crystal Grove mine, near Lasellsville, NY. Pockets from 3 to 15 centimeters in diameter. May 30, 2011.*

present themselves nicely on sparkling beds of these crystals instead of the usual bare dolostone seen from most of the other locations in the region. The crystals in these pockets can form in any orientation within the pocket, including the walls and ceiling. Attractive, damage-free specimens are common but almost always small.

Occasionally, larger pockets to over a meter in diameter are encountered. These are exceptional, rare finds because they provide wonderful matrix specimens as found in the smaller pockets, but with larger crystals sometimes exceeding 15 centimeters. Larger crystals will normally be detached from the wall rock and often found damaged as a result. Clusters are sometimes present, yet almost always need to be reassembled. Floaters are uncommon at Crystal Grove and oddities like scepters, enhydros, and skeletal crystals even rarer. Also rare are any associated minerals other than crystallized dolomite. When encountered, the dolomite crystals are usually less than a centimeter in length and are normally sharp and lustrous. Occasionally, they appear on broken surfaces of dolostone wall rock that has collapsed off of the pocket's inner surfaces, showing that their formation was sometimes very late in the crystallization sequence of events. Calcite is occasionally present in disk-shaped or blade-shaped rhombohedral crystals and is almost always weathered. Hydrocarbons, relatively common at other locations in rich black flakes, are considered an oddity when found here. The hydrocarbon is also normally weathered to a dull gray color unless it is preserved as inclusions within the quartz.

This fee location is well suited to beginners and provides a maintained campground with cabins, showers, and recreational areas. The shop has supplies and specimens for those who do not find enough to make them happy or those who prefer to purchase their specimens

Fig. 20a. *Typical small specimens on druse. Crystal Grove mine, near Lasellsville, NY. Above, 5.5 cm; below, 3 cm. Collected May 28, 2011.*

. No claims were allowed here at the time this was written, so the serious Herkimer diamond miners who spend weeks at a time collecting seldom frequent the Crystal Grove diggings. The location is the perfect fit for families and collectors not ready for working walls.

Fig. 20b. *Typical small crystals on druse. Crystal Grove mine, near Lasselsville, NY. 4.8 cm. Collected May 28, 2011.*

Fig. 21. *One of several collecting areas. Crystal Grove mine, near Lassellsville, NY. Spring, 2012.*

Fig. 22. Left: *Just-opened pocket releasing fog. Crystal Grove mine, near Lassellsville, NY. Field of view, 25 cm. May 23, 2011.* Right: *Lower section of the same pocket 10 minutes later. Both views show quartz crystals mixed with broken dolostone.*

Fig. 23. *Best matrix specimen from previous pocket (Fig 22). Nice transparent crystals on druse-covered matrix. Crystal Grove mine, near Lassellsville, NY. 6.5 cm. Collected May 23, 2011. Jay Walter collection.*

Fig. 24. *An unusually large single crystal on druse-coated matrix. Crystal Grove mine, near Lassellsville, NY. 11.5 cm with the largest crystal measuring 5.7 cm. Collected May 23, 2011.*

Diamond Acres, Mohawk Township, Montgomery County — commonly referred to as the Margaret Hastings property, Hastings farm, or simply Fonda — is also operated as a fee digging site found in the Little Falls Dolostone. This location is a property which remains forested, although it has been heavily excavated. Day diggers and beginners are welcome, but many who dig at Diamond Acres have their own claims. These are marked out on the surface of the landscape with ropes or caution tape for boundaries, and trails connect the labyrinth of private digging sites. Diamond Acres is currently the only site that allows power tools to be used to collect specimens. Day visitors have to find spots between claims to dig and opportunities to do this exist in the soil as well as the ledge rock. One of the more unique aspects of this location is how claim holders "decorate" their digging sites. Stone walls made from waste rock are used in many places to construct divisions, steps, rooms, fire pits, and any number of other features. At times, the location has the appearance of a medieval village and the miners often look the part! The man-made vertical relief on this flat section of forest floor is dramatic. A great effort has been spent on preserving the forest so that this location does not become an open pit quarry and diggers will still be able to find shaded digging areas. Many claim owners are willing to show beginners how to find crystals. Primitive camping and bathroom facilities can be found on site.

The primary dome pockets occur in similar size to those found in Middleville, sometimes even larger, but they are not in such a well-defined layer. They vary more in their positions within a thick layer of table rock and are not quite as predictable to find as those in Middleville. Approximately

Fig. 25. *Cluster of two large crystals. Diamond Acres mine, Mohawk Township, NY. 9 cm. Collected 1994. Jay Walter collection.*

Fig. 26. Emptied pockets. Diamond Acres mine, Mohawk Township, NY. September 4, 2011. Pockets to 60 centimeters.

0.5 to 1 meter below these, sub-pockets and jewelry pockets can be found. Like the other locations that have sub-pocket layers, these openings often produce higher quality, but smaller, crystals. Here, they commonly contain anthraxolite.

It was here that the largest recorded Herkimer diamond was recovered by Lt. Bill Francis and reported in *Rocks and Minerals* in 1953. It weighed 8.1 kilograms and was 22 x 18 centimeters in size. Big crystals are not an oddity here and many other huge Herkimer diamonds have been recovered from this section of land. Many pockets have produced large crystals, but larger is not always better because these larger Herkimer diamonds tend to have greater concentrations of internal fractures and are often shattered, probably because of frost action and, perhaps, tectonic activity. The crystals are rarely attached to the matrix, and clusters almost always need reconstruction, as is the norm at most of the region's sites. Blowouts of inclusions, causing shallow chips off the crystal's surface, also tend to be very common in these large primary pockets, indicating that there were at one time liquid-filled inclusions (enhydros) in many of these specimens that later froze, expanded, and broke the crystal. Associated minerals other than the occasional poor quality calcite are very rare. The wall rock is almost always barren in the largest pockets.

Fig. 27. Open pocket with Herkimer diamonds in situ. Diamond Acres mine, Mohawk Township, NY. Field of view, 30 cm wide. September 4, 2011.

Fig. 28. Skeletal Herkimer diamond with minor clay inclusions. Diamond Acres mine, Mohawk Township, NY. 12 cm. Previously in Ernie and Vera Schlichter collection.

Hickory Hill is an active fee location near Fonda in the Mohawk township of Montgomery County. This Little Falls Dolostone exposure was previously known as the Barker farm. The owners only open this commercial operation on holiday weekends three times a year, yet attract diggers from all over North America.

The collecting areas are on a section of land adjacent to a large farm field where a number of diggings have been established among the hardwood trees in the forest. None of the diggings are deep; rather, the areas are scattered, shallow, and allow for liberal choices when it comes to the methods one wants to employ to find crystals. Sifting virgin soil, lifting weathered rock layers with bars, or working more traditional table rock can all be done on this property. Often collectors will simply walk the open areas and visually locate crystals on the soil's surface that have weathered free or been exposed by the owner's frequent bulldozing work.

The specimens seem to be quite consistent in form. Most are beautiful, euhedral, single crystals or small clusters found in small pockets seldom over 10 centimeters in size.

Weathered pockets are larger, but the crystals that they contain — water clear gems that form without any points of attachment — are classic Herkimer diamonds in all respects.

In the location's history, there have been any number of unusual finds of significance. Specimens such as nice skeletal crystals and scepters have been found of a quality equal to the more frequented commercial sites. Occasionally, pockets very rich in massive hydrocarbon are encountered that contain Herkimer diamonds with interesting and unusual hydrocarbon contact surfaces. Because the area available does cover many acres, there is always the potential for finding oddities that are not just classic Herkimer diamonds.

Hickory Hill is well suited for beginners, children, and more serious collectors wanting to break rock. Shade is ample and primitive field camping is available on site.

Fig. 29. *One of many collecting areas along forest's edge. Hickory Hill mine, Mohawk Township, NY. Spring 2011.*

Fig. 30. *Open weathered pocket. Hickory Hill mine, Mohawk Township, NY. Collected Spring 2011.*

Fig. 31. *Perfectly clear Herkimer diamond. Hickory Hill mine, Mohawk Township, NY. Approximately 2.8 cm. Collected by Bill and Dianne Petronis. Bill and Dianne Petronis collection.*

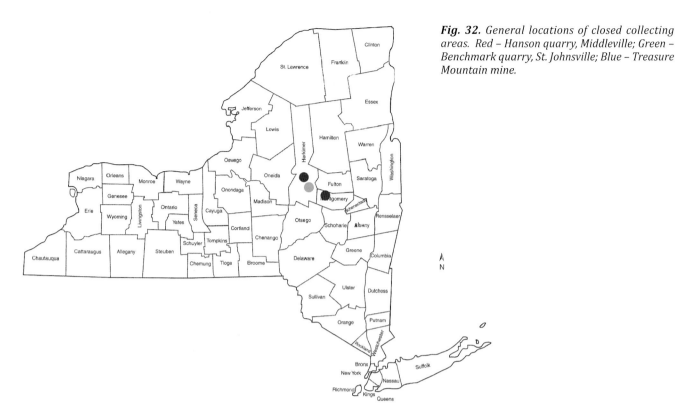

Fig. 32. *General locations of closed collecting areas. Red – Hanson quarry, Middleville; Green – Benchmark quarry, St. Johnsville; Blue – Treasure Mountain mine.*

Closed Locations

The **Benchmark quarry** near St. Johnsville in Montgomery County is an active dolomite quarry in the Little Falls Dolostone and is owned and operated by Hanson Industries. Previously, the location has been referred to as the Talarico or Eastern Rock Products quarry or just the St. Johnsville quarry. Currently, it is strictly off-limits to collectors. In the past, club outings were sanctioned into the quarry, but it has been several years since that option has been available to collectors. As a result, there have been many who have trespassed onto the property, mostly at night, to collect. Others have even gone to the extent of rappelling into the quarry to gain access. Trespassing or unauthorized entrance onto any property is strongly discouraged. These practices contribute to the lack of success that clubs and legitimate collectors have in gaining permission to collect on these sites.

The pockets found in this quarry can attain enormous size when compared to other locations. Openings to several meters, in some cases interconnected pockets, have been found. As a result, the minerals recovered from them have the potential to attain large size. Further adding to the appeal of large pockets and large crystals is the fact that probably the greatest diversity of mineral species and crystal oddities have come from this location. There are currently three horizons showing in the exposed dolostone along which Herkimer diamond pockets have formed.

The Herkimer diamonds from the Benchmark quarry can exceed 15 centimeters in diameter and are occasionally found in clusters over 35 centimeters. Matrix specimens are as common as floater crystals and the matrix on which they form is often coated by well-formed white to cream-colored dolomite crystals. While large crystals are relatively common, small gemmy crystals, so common from other locations, are less so. Skeletal crystals are present, but are seldom as well developed as those found at other locations, such as Middleville and Fall Hill. Large crystals are often smoky and will sometimes have phantoms. Both two- and three-phase inclusions are common: two-phase inclusions have water with gas bubbles (often movable) while three-phase inclusions have water, gas and solids, usually hydrocarbons, within the sealed inclusion.

Black-stemmed scepter crystals are a special form found at the St. Johnsville quarry. These crystals exhibit a traditional Herkimer diamond crystal on the top of a thinner, elongated crystal, thereby resembling the scepters of medieval royalty. The shaft is often filled by dark black hydrocarbons, creating a pleasing contrast. The different sections of the crystal are normally in parallel orientation. Scepters appear to result from two phases of growth in slightly different environmental conditions. The shaft would grow first and the cap, later. The shafts of St. Johnsville scepters are so highly included with hydrocarbons that they tend to appear completely opaque.

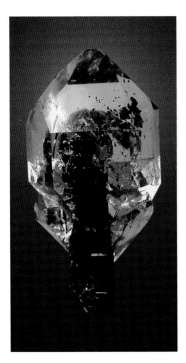

Fig. 33. *Scepters. Benchmark quarry, St. Johnsville, NY. Left: 2.5 cm. Right: 4 cm. Stuart Strife collection.*

Appendix D shows diagrams of typical scepters. The ones from here are almost always singly-terminated and were once attached to matrix. They occur in a mineralized band in a particular layer about a meter thick that cuts diagonally across the quarry about 30 meters from the present quarry floor. The pockets in this layer are small and contain abundant hydrocarbon that is included in the dolomite, calcite, and scepter stems. Later calcite and the tips of quartz scepters are often free of hydrocarbon. Scepters from St. Johnsville are exceedingly rare and highly desired by collectors.

Attractive matrix and floater Herkimer diamond specimens are sometimes further enhanced by what are universally agreed to be the finest calcites to come from the Little Falls Dolostone. Good quality calcite specimens are common and crystals reaching a whopping 15 cm in diameter exist. Unlike the calcites from other locations that are almost always weathered due to groundwater flow, the calcites recovered from the rock at the Benchmark quarry are almost always pristine. Normally, they present themselves as disk-shaped rhombohedral crystals of a honey yellow to amber color. They are often nearly or completely transparent and generally intergrown with the Herkimer diamonds with which they formed in these large pockets. Calcite appears to form before, at the same time as, or after the quartz, depending on the pocket. The paragenesis of the quartz and calcite here shows a greater overlap than at other locations. Inside these calcites one can often find acicular needles of the mineral marcasite. This sulfide grew before the calcite and the calcite then formed, encasing the fragile crystals. Sometimes the marcasite is not fully encased and extends out of the surfaces of the crystals.

Fig. 34. *Calcites on dolomite. Benchmark quarry, St. Johnsville, NY. 8.9 cm.*

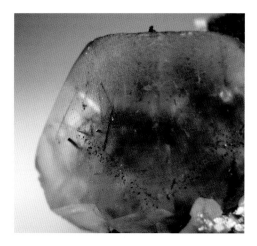

Fig. 35. *Marcasite inclusions in calcite. Benchmark quarry, St. Johnsville, NY. Field of view, 6 cm.*

Fig. 36. *Calcites with pyrite and dolomite. Benchmark quarry, St. Johnsville, NY. 5.5 cm.*

Fig. 37. *Atypical calcite on dolomite, with marcasite inclusions. Benchmark quarry, St. Johnsville, NY. 9.2 cm.*

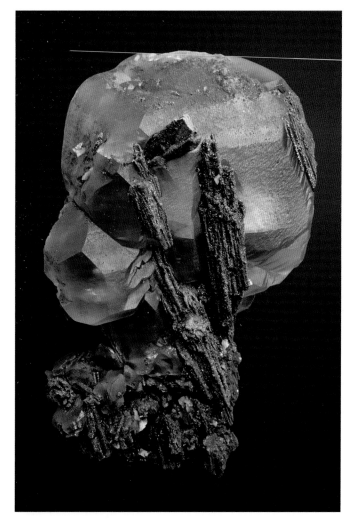

The other iron sulfide, pyrite, is commonly found in association with both the quartz and calcite. Pyrite occurs as rods, stalactitic forms, draperies, and occasionally as crystals both inside calcite and quartz or, like with marcasite, not completely encased by them.

One example of a pocket containing small crystals of sphalerite was also discovered around 2004.

The **Hanson quarry at Middleville**, also commonly known as the Middleville quarry, is another Little Falls Dolostone site. Until the late 1960s, the quarry was abandoned and routine

Fig. 38. *Exceptional calcites with pyrite and dolomite. Benchmark quarry, St. Johnsville, NY. 23 cm.*

Fig. 39. *Pair of large Herkimer diamonds on dolomite matrix. Benchmark quarry, St. Johnsville, NY. 13 cm with crystals measuring 10 cm each.*

Fig. 40. *Calcite, Herkimer diamond, marcasite, and dolomite. Benchmark quarry, St. Johnsville, NY. 15 cm.*

Fig. 41. *Dolomite on sphalerite crystal. Benchmark quarry, St. Johnsville, NY. 2.5 cm.*

Fig. 42. *Herkimer diamond on calcite. Hanson quarry, Middleville, NY. 2.9 cm. Collected by Bill and Viki Hladysz. Bill and Viki Hladysz collection.*

collecting, mostly by locals, was widespread. Eastern Rock Products purchased the quarry and reopened it in 1988 to meet the local need for crushed rock for road improvements. Collecting continued until 1995, at which time the quarry owners restricted these efforts and patrol of the property began.

The quarry produced traditional Herkimer diamonds from its lower bench level to 15 centimeters in size, although most were considerably smaller. On this same level, small calcites of numerous colors were commonly found as coatings on the quartz. Most of the best specimens were thumbnail-sized (less than 3 centimeters) examples. Domed mud-filled pockets,

however, were encountered rarely. Varieties such as skeletal and scepter crystals are known but are also exceedingly rare.

The upper bench in the quarry was not a producer of quartz, but was well known for producing a broad diversity of interesting calcite specimens. Sprays of steep rhombohedral crystals in attractive groups to 10 centimeters, flower-shaped clusters, mammillary forms, and parallel growth groupings are among some of the varieties found. Calcite crystals were yellow, gray, white, red, green or pink. Dolomite in curved crystals was also plentiful in gray, pink, and creamy white. Sphalerite was also rarely found in crystals to 1.5 centimeters.

Fig. 43. *Calcite. Hanson quarry, Middleville, NY. Crystal cluster 2 cm. Collected by Bill and Viki Hladysz. Bill and Viki Hladysz collection.*

Fig. 44. *Herkimer diamond in calcite. Hanson quarry, Middleville, NY. 9 cm. Collected by Bill and Viki Hladysz. Bill and Viki Hladysz collection.*

Fig. 45. Sphalerite. Hanson quarry, Middleville, NY. 1.5 cm. Collected by Bill and Viki Hladysz. Bill and Viki Hladysz collection.

Historically, crystals have also been reported from the vicinity of Newport, north of the Middleville quarry and from various sites near Salisbury and Salisbury Center; e.g. Diamond Ledge on Ives Road operated as a fee locality more than thirty years ago by George and Rachel Moholsky, and Diamond Hill. Herkimer diamonds have also been collected from Crane's Hollow near Amsterdam.

Fall Hill is a rise along the south side of the Mohawk River in the town of Little Falls, Herkimer County. Fall Hill contains a remarkable exposure of approximately 120 vertical meters of the Little Falls Dolostone. Within these layers of rock, at least seven different horizons containing Herkimer diamonds have been discovered. Exposures extend from Finck's (or Fink) Basin area westward for several kilometers. The area was worked by many different locals for decades, yet did not become commercialized until the late 1990s because all the productive outcrops were on private lands.

In the late 1990s, part of Fall Hill was purchased and opened as a commercial digging locality named the **Treasure Mountain mine**. The history of this site and the collection conducted during its operation from 1999 and 2003 are well documented in an article for *Rocks & Minerals* magazine. The authors, Borofsky, Whitmore, and Chamberlain, describe the details of the site's opening to the public and gave descriptions of the fine minerals found there. A segment of the PBS program *Real Science* was taped at the location in 1999 and a *Rock and Gem* magazine article by M. Walter followed in 2002. The site was visited by thousands and appeared to be taking shape as a popular tourist destination for the future.

Collecting there ceased when the ownership of the property changed. Collecting is no longer allowed. The

Fig. 46. Top: Public digging area looking east. Treasure Mountain mine, Fall Hill, Little Falls, NY. Summer 2001. Bottom: Five freshly opened pockets (yellow arrows). The fifth is barely visible on the far left of the photograph. Treasure Mountain mine. Summer 2002.

specimens, however, are still seen in collections and museums, and many rank as the most unique and aesthetic quartz specimens ever produced in New York State.

Treasure Mountain pockets are small when compared to the majority of other locations. Openings of 30 centimeters are large, although occasional openings exceeding a meter in length have been found. Most common are vugs of 5 to 15 centimeters in diameter. There was not a well-defined horizon for pockets, but they were generally found in a horizon measuring about one meter thick in the public digging area. The area open to the public was about 150 meters in length and provided an excellent exposure to this rock layer in the form of a short wall that seldom exceeded three meters in height.

The majority of the excitement generated during Treasure Mountain's operation centered on the sceptered crystals found there. Although very rare at all other collecting areas, Herkimer diamond scepters were not unusual at Treasure Mountain. Normally small, this rare form of Herkimer diamond could reach 15 centimeters in length. Unlike the singly-terminated scepters from the Benchmark quarry, they

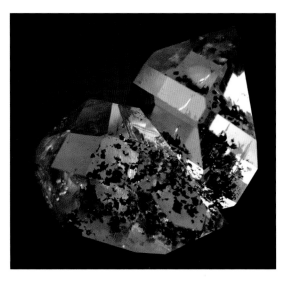

Fig. 47. *Typical scepter. Treasure Mountain mine, Fall Hill, Little Falls, NY. Collected Summer 2001. 4.8 cm.*

Fig. 48. *Skeletal crystal. Treasure Mountain mine, Fall Hill, Little Falls, NY. Collected Summer 2001. 4.9 cm.*

Fig. 49. *Hydrocarbon-included Herkimer diamonds. Treasure Mountain mine, Fall Hill, Little Falls, NY. Collected Summer 2001. 4 cm.*

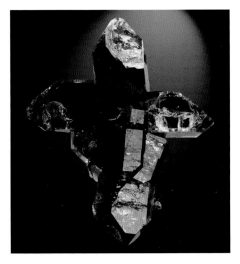

Fig. 50. *Herkimer diamond cross. Treasure Mountain mine, Fall Hill, Little Falls, NY. 5 cm.*

Fig. 51. *Black crystal cluster. Treasure Mountain mine, Little Falls, NY. 2 cm.*

Fig. 52. *Gemmy clear Herkimer diamonds in calcite. Treasure Mountain mine, Little Falls, NY. 4.6 cm. Jay Walter collection.*

more commonly formed as floaters or as doubly-terminated crystals locally called dumbbells or barbells. This variety of scepter shows clear quartz overgrowth on both ends of an elongated, translucent, black stem instead of on one end only. Although most scepters from Treasure Mountain are floaters with no points of attachment, some can be found attached to a matrix of cream to tan colored calcite, or crystallized dolomite. The calcites, though often well formed, are almost always slightly weathered. Further, they often have secondary calcite overgrowths that detract from their appearance. These overgrowths both fluoresce and phosphoresce while the underlying calcite does not. The diversity of Herkimer diamond forms, associations, and sizes found here is similar to the diversity of specimens from the Benchmark quarry.

Other oddities like inclusions of negative crystals, pyrite rods, anthraxolite, and enhydros, as well as phantoms, crosses, and black crystals were commonly encountered. Skeletal crystals are in high concentration here, with one to five centimeter examples being common. Large skeletal specimens can exceed 12 cm in diameter.

Although short-lived, the Treasure Mountain mine made a major impact on the mineral collecting world. Currently closed, it can only be hoped that this property will again open to the public at some time in the future.

  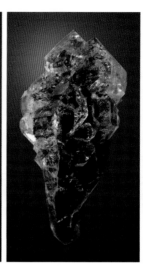 

Fig. 53. *Scepter with calcite. Treasure Mountain mine, Fall Hill, Little Falls, NY. 9.5 cm. Collected by John Kulla and Chris Phetteplace. Previously in the John Kulla collection.*

Fig. 54. *Scepters. Treasure Mountain mine, Fall Hill, Little Falls, NY. Left to right 6.3 cm, 6.5 cm, and 5.8 cm. All collected by Paul Nuckols.*

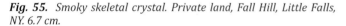

Fig. 55. *Smoky skeletal crystal. Private land, Fall Hill, Little Falls, NY. 6.7 cm.*

Fig. 56. *Complex scepter cluster. Private land, north of Little Falls, NY. 10 cm. Collected in 2006 by John Kulla. Previously in the John Kulla collection.*

Other collecting sites on private lands have been exploited for Herkimer diamonds on Fall Hill over the decades. These sites also have scepter horizons and occasionally produce noteworthy crystals.

It has long been known that nice Herkimer diamonds, including fine scepters, can also be found on the opposite side of the Mohawk River north of the town of Little Falls. All collecting that has been conducted in this region has been done on private lands, none of which have been open to the public. Those planning to dig for specimens in this area need to do careful research and make arrangements with the landowners where specimens can be found.

Locations Outside the Traditional Herkimer Diamond Region

The Herkimer diamond region has not normally included the locations featured in this section. Although traditionalists would not consider these areas to have "Herkimer" diamonds, the quartz crystals are for all intents and purposes identical to those found in the Herkimer diamond region in two ways: 1) They occur in corresponding sedimentary layers of rock and are not the result of formation in igneous rocks or metamorphic formations in the surrounding areas (which also occasionally contain quartz that is visually similar to the Herkimer diamonds), and 2) They appear to have formed by the same geological mechanisms that produced other Herkimer diamonds. So for convenience of discussion, the crystals from these peripheral areas of New York State and southern Ontario and Quebec, Canada, will be referred to as Herkimer-style quartz.

In the early 1950s, the **Gailor Stone quarry** in the town of Saratoga Springs, Saratoga County, was producing interesting Herkimer-style quartz crystals and associated minerals. This long closed locality is described in Rowley's paper (1951) as having pockets usually in the 30-centimeter range that contained Herkimer diamonds always found in association with dolomite and/or calcite. The specimens were almost always found on matrix and of the quality of the best material that was coming from Little Falls at the time. Rowley gives great detail regarding a meter long pocket that was found to contain large crystal fragments that showed dissolution while other fragments within the same opening showed regrowth. Both characteristics are very rare, or unheard of, at other Herkimer diamond localities. Other oddities noted include cubic pyrite crystals, sphalerite, cerussite, aragonite, and small Herkimer diamond scepters.

Diamond Point and **Diamond Island** on Lake George, Warren County (Lake George diamonds), are outliers of the Beekmantown group. Single Herkimer diamond-style quartz crystals have been collected in the past along the shoreline. These crystals are relatively transparent, often with slightly rounded edges from mechanical weathering. They typically range from less than 0.6 to more than 11 centimeters. No sites for *in situ* collecting are known, so matrix specimens are likely exceedingly rare. Both sites are privately owned and no collecting has been possible for many years.

The **Ogdensburg Stone quarry**, another Beekmantown group location in northern St. Lawrence County, has produced Herkimer-style quartz crystals. Known specimens seem to have been produced in limited numbers during a short period of time in the 1990s. Herkimer-style crystals in small gemmy specimens with dolomite were recovered along with crystals of a sulfate, either barite or celestine.

Fig. 57. *General locations of localities outside the traditional Herkimer diamond region. Magenta – Ogdensburg quarry; Green – Norwood quarry; Blue – Diamond Point & Island; Red – Gailor Stone quarry; Yellow – Lowville occurrence.*

Fig. 58. *Herkimer-style quartz. Gailor Stone quarry, Saratoga Springs, NY. 4.9 cm. NY State Museum specimen, #11218. Ex. Elmer Rowley collection. Stephen Nightingale photo.*

Fig. 59. Herkimer-style quartz crystals. Lake George, NY. Largest crystal 1.3 cm. NY State Museum specimen #11357. Ex. Elmer Rowley collection. Stephen Nightingale photo.

Fig. 60. Herkimer-style quartz. Lake George, NY. 4.5 cm. NY State Museum #22367. Ex. Philadelphia Academy of Sciences. Stephen Nightingale photo.

The **Barrett quarry** near the town of Norwood, St. Lawrence County, is probably the most northerly location in the state that has been known to produce Herkimer-style quartz with any frequency. It operates in a part of the Beekmantown Group that is younger in age — lower to middle Ordovician — than the Little Falls Formation. Better known for its attractive pink to cream colored dolomite crystals, this active quarry produced fine specimens, including Herkimer-style quartz crystals, during the 1970s and 1980s. The pocket layer has been left exposed in an upper level of the quarry and collectors no longer have access to specimens from this site. It is a distinct layer about midway between the quarry floor and road level. The pockets are usually between 10 and 50 centimeters in diameter. Herkimer-style quartz of traditional form with moderate internal fracturing and up to five centimeters in diameter has been found loose and on matrix within the pockets. The most common mineral association is calcite, which forms in attractive crystals of moderate luster in both schalenohedral and rhombohedral crystals.

This quarry does not allow collecting and it is unlikely that specimens would be produced even if it were. All quarry operations are now occurring on levels well below where the pockets were found, and there has been no indication of pocket-containing layers on levels now being actively worked.

The Roundout Formation, of upper Silurian and lower Devonian age, contains dolostones with occasional cavities containing Herkimer-like quartz crystals, often with large specimens similar to those sometimes found at Diamond Acres in Fonda. This dolostone contains enough clay to be used as hydraulic cement and commercial mining at the **Rosendale deposit**, near Kingston in Ulster County, which produced numerous fine quartz specimens, a suite of which is preserved at the New York State Museum.

A new locality simply called **Lowville** has recently been producing specimens seen for sale on the mineral market. The location and details of the find have been kept secret because the commercial mining of specimens is on private land somewhere near Lowville in Lewis County. Information provided indicates that the occurrence is in the Beekmantown Dolostone of early Ordovician age. Several pockets to half a meter in diameter have produced specimens that include Herkimer-style quartz to eight centimeters with noteworthy variety: skeletal crystals, scepters, phantoms, prismatic crystals, and black crystals have all been found. It appears that the crystals that exceed several centimeters are nearly always skeletal in form and usually pale to dark brown in color. Small crystals are virtually indistinguishable from the small crystals found at the localities within the traditional Herkimer diamond region. Crystals can be found on matrix and as floaters, free of any attachment to rock.

The site has also produced diverse forms of calcite, including modified rhombohedral crystals, well developed nailhead forms, scepters, and complex phantoms similar to the material produced from the Grant quarry in Ontario, Canada, and found within the same-aged layer of dolostone. The only recognized twinned calcites of any Herkimer diamond or Herkimer diamond-like occurrence have been discovered here. Two generations of calcite growth are evident on many specimens. The first generation is often composed of schalenohedral forms, with the second generation being transparent rhombs. This sequence leads to rhombehedral-shaped crystals with schalenohedral phantoms or scepters with scalenohedral stems and rhombohedral tips. In some cases, the calcite crystals form after the development of the quartz, as would be expected. In many cases, however, the calcites have begun their growth during the late phases of quartz crystallization, leading to calcites partially embedded in the quartz crystal faces. The calcite crystals range from colorless to cream-

Fig. 61. *Calcite and dolomite. Barrett quarry, near Norwood, NY. 8 cm. Collected by Vern Sawyer, 1960. Steven C. Chamberlain Collection, #12279 and photo.*

Fig. 62. *Herkimer-style quartz on dolomite and calcite. Barrett quarry, near Norwood, NY. 7.7 cm. NY State Museum specimen, donated by Dr. Steve Chamberlain, # C3308. Stephen Nightingale photo.*

Fig. 63. *Smoky Herkimer-style quartz with calcite. Lowville, NY. 4 cm. Mizutani /Nakanishi collection specimen and photo.*

Fig. 64. *Smoky, skeletal, Herkimer-style quartz with yellow and red calcite. Lowville, NY. 5.5 cm.*

Fig. 65. *Very large rhombohedral calcite with brown hydrocarbon inclusions. Lowville, NY. 9.3 cm.*

Fig. 66. *Calcite and pyrite on Herkimer-style quartz with pyrite inclusions. Calcites show nice hydrocarbon and sulfide phantoms. Lowville, NY. 4.2 cm.*

colored, with yellows, ambers, and reds being equally common. They are seldom over three centimeters in length and often transparent and with exceptional luster. Pyrite is also a common mineral at this location and is found as inclusions in crystals, coatings on all the minerals previously mentioned, and in plates of crystals of unusual form and black color to eight millimeters in diameter, with second generation pyrite on some specimens. Late stage pyrite has also been found in skeletal faces of quartz crystals and on broken quartz crystal surfaces, including scepter stems.

Specimens that are formed from balanced combinations of numerous species, including quartz, calcite, dolomite, and pyrite make for attractive specimens. Such combination specimens are common from this location and are some of the most desirable specimens available today.

Fig. 67. *Calcite and pyrite on skeletal Herkimer-style quartz cluster. Lowville, NY. 8.1 cm.*

Fig. 69. *Two generations of calcite growth. The calcites exhibit scepter-like and complete overgrowth of rhombohedral forms over schalenohedral forms. Grant quarry, near Greely, Ontario, Canada. Left: 2 cm. Right: 1.5 cm. Photos by George Robinson reproduced with permission of the Canadian Museum of Nature, Ottawa, Canada.*

Fig. 68. *Quartz on calcite and dolomite. Grant quarry, near Greely, Ontario, Canada. 8.1 cm.*

Canadian Locations

Grant quarry, near Greely, Ontario, Canada, is an active quarry in the Beekmantown Group. Owned by Peter and David Grant of Cornwall, Ontario, the quarry was opened in 1987 and began producing quality mineral specimens within several years. The quartz is normally small and clear, identical to the New York State Herkimer diamond, or larger and smoky. Black crystals are common. On very rare occasions, scepters have been noted. Skeletal crystals are common and specimens form both on and off of the dolostone matrix.

The calcites have interesting secondary rhombohedral forms as overgrowths on schalenohedral forms. No fewer than 18 forms have been documented (Robinson et al., 2011) and the crystals are nearly identical to those found in Lowville, New York. The colors, sulfide and hydrocarbon inclusions, and other characteristics add to the uniqueness of this material.

Blair Hill farm near Greely, Ontario, Canada, is also in the Beekmantown Group and began producing Herkimer-style quartz and calcite in 1968. A nearby excavation for a natural-gas pipeline revealed quartz crystals, and soon thereafter local mineral collectors began active work on the property. Open collecting continued for a decade until a local mineral collector leased the property. This period of open collecting produced many fine specimens. In size and form the Herkimer-style quartz found at the Blair farm is nearly identical to the material found at the Grant quarry. The calcites, however, are highly weathered and not as fine as those from the Grant quarry.

Fig. 70. *Smoky, partial scepter. Blair Hill farm, Greely, Ontario, Canada. 5.3 cm. Previously in the George Robinson collection.*

Other localities in the Beekmantown Group have also produced Herkimer-style quartz specimens, but in smaller quantities and with less desirability. Some of those locations include the **Tomlinson quarry**, Kenmore Corners, Ontario, Canada; the **Forbes quarry (Merrickville quarry)**, Merrickville, Ontario, Canada; the **Maple Grove quarry**, Kemptville, Ontario, Canada; and the **Marcil (LaFarge) quarry**, near Ste-Clotilde-de-Chateauguay, Québec, Canada.

Chapter 5:
The Gear and How to Use It

If you are ready to field-collect some Herkimer diamonds of your own, what equipment will you need? If you plan a simple family outing with the kids, you may elect to rent tools at the site and, therefore, need hardly any specialized gear of your own. If you are setting up camp for the summer or intend to bring down a high wall of rock, you will need a different approach altogether, as discussed below.

Basic Gear

First consider a simple day trip with youngsters. Some of the commercial locations you would be likely to visit are happy to rent the tools you may need for a reasonable fee. These will most likely be crack hammers and chisels designed for breaking open small rocks. In that case, just have your old work clothes, some sturdy shoes or boots, and you will be on your way. Other supplies that will come in handy are newspaper for wrapping specimens, a bucket to carry your gear, and work gloves. To dig in the soil for crystals, common garden tools like shovels, trowels, and cultivators work well.

Mining Herkimer diamonds by digging for pockets and breaking down wall requires specialized tools. Here is a partial list:

Gloves
Steel-toed boots
Safety glasses
Crack hammers
Sledgehammers
Cold chisels of various sizes
Screwdrivers
Whisk broom
Small pry or gad bars
Large 6-foot plus crowbars
Wrapping and packing materials
Extra long tweezers
Shovels
Sifter
Bucket with water (for cooling crystals if digging during the spring)
Tarps, rope, and poles (for creating shade if in direct sunlight)
First aid kit
Wood-splitting wedges
Plate steel
Capeing tools
Spring-steel wedges

Safety is always a concern and appropriate precautions should be observed. Gloves to protect your hands and safety glasses to protect your eyes are musts. Also recommended are steel-toed boots. Having a first aid kit on hand can also come in handy for dealing with minor accidents. The author and publishers assume no liability for failures of design, improper manufacture, or misuse of these or any other tools described herein.

Fig. 71. *A few of the basic tools.*

Hammers

Be aware that hammers made to break rock are not the same thing as carpenters' claw hammers or ball – peen hammers, which are worthless for breaking rock or for hitting the chisels used for mineral collecting. Rock hammers are called crack hammers and usually come in two- and four-pound sizes. Larger hammers are normally called sledgehammers and are available in 8-, 12-, and 16-pound sizes. Exceptionally large hammers of 20 pounds or more are available through specialized tool suppliers. It is difficult to get sledgehammers in excess of 20 pounds in the United States.

Big hammers — sledges — are easier to handle if their handles are wrapped to help increase the friction between your hands and the tool. This holds especially true when working in wet conditions. The following series of photographs shows one method of handle wrapping using duct tape that is similar in style to how hockey players wrap their sticks. Other varieties of tape can be used for this purpose.

Fig. 72. A more advanced group of tools.

Fig. 73. Some safety gear including steel-toed boots, gloves, safety glasses, and a first aid kit.

Fig. 74. Hammers: 20 lbs, 16 lbs, 8 lbs, 4 lbs, 2 lbs.

A tape job such as this should be completed with a single length of duct tape, never having to tear it until the end of the wrapping is reached. The addition of a hammer handle protector is a good idea as well. Some people also paint their tools a bright color so they are easier to find when it is time to leave and even stamp or engrave an identifying mark, name, or contact information into the tool.

Fig. 75. *Step 1. Overlay tape on butt of handle and diagonally wrap handle.*

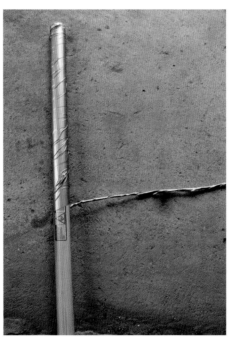

Fig. 76. *Step 2. "Spin" the tape making a rope to wrap the area taped in step 1.*

Fig. 77. *Step 3. Diagonally wrap the rope around the handle.*

Fig. 78. *Step 4. Do several wraps on the butt end of the handle.*

Fig. 79. *Step 5. Diagonally wrap the length, again.*

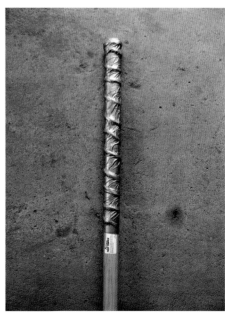

Fig. 80. *Step 6. Tear off tape and press down the entire length of your completed work.*

Fig. 81. *Hammer handle protectors.*

Fig. 82. *Dennis Snyder swinging from the side.*

Fig. 83. *Jay Walter swinging between his legs.*

Swinging large sledgehammers in tight spaces and for extended periods of time takes some practice. The techniques vary among collectors, but two methods are often employed more than others. The first is swinging and hitting steel between your legs. The second involves attack from the side. For both methods, the hammer should be lifted as high as possible while still maintaining good control and accuracy. In lifting, your legs and arms should be used as much as possible. Muscles in the back and abdomen can be easily strained. Arms will get tired, but seldom strained to the point of failure. Steel-toed and steel-shanked boots are a real plus here as one misfire will prove.

Chisels

Quality cold chisels that will stand up to the abuse of breaking dolostone are uncommon. You will need to experiment with different brands to find ones that perform to your expectations. Cheap chisels will normally not do a good job. It is sometimes helpful to have various lengths on hand. Some varieties will come with a plastic protector for your hand. Expect to break your fair share of chisels even if you use them correctly.

As you progress as a collector, you are likely to find that you will be using the next tool, spring-steel wedges, more and

more and chisels, less and less. Eventually, chisels will likely find use in cobbing down matrix and in pocket excavations only.

Spring-Steel Wedges

Spring-steel wedges are made from the old leaf springs of automobiles; sometimes, in extreme cases the tines from a fork lift are used. A length of the spring is cut to the desired size and then a sharp edge is ground down on one end, leaving the opposite end blunt. Spring-steel has a temper that should be retained at all times during the process of making it into wedges, as it is critical to the tool's performance when in use. Deviation from the established construction practices described here will put the user and others at risk.

The cutting of the steel should be done using a steel band saw or radial arm saw designed for working with steel or, at home, with a special carbide steel-cutting circular saw blade. Cooling the metal is important so it does not lose its temper. It should never get red hot at any phase of making the wedge. Although it is unlikely to overheat at this stage, care should be taken to cool the steel using water or oil. If the temper is lost on this cut edge, it will mushroom excessively during use. If the steel is too hard with too much temper, the risk of having shards of

Fig. 84. *Raw spring steel.*

Fig. 85. *Mixed spring-steel wedges.*

Fig. 86. *Short vs. long steel.*

Fig. 87. *Flat points.*

Fig. 88. *Tapered points.*

Fig. 89. *Scalloped points.*

steel splay off when pounded on with a sledgehammer increases. Sharp edges can be rounded using a grinder or sander if desired.

The dangers associated with not properly tempering spring-steel cannot be overstated. Flying shards of steel behave like bomb shrapnel or bullets. Serious injury or death can occur! The cut end should be the end that is hammered on when in use and the opposite end, which is likely already partially tapered, should be ground down into an edge.

The grinding phase involves fashioning an edge onto the end of the steel that will be pounded into the rock. Standard shop bench grinders can be used for this purpose. Once again, care must be taken not to lose the temper of the spring-steel. Constant cooling should occur as the edge is created. Never grind for more than a few seconds without bringing the temperature of the steel back down. This is usually accomplished by having a bucket of water next to the grinder and dipping the steel in it after each pass across the grinder. Doing several pieces of steel as a group allows for proper cooling time. For example, have five pieces of steel — give one a single pass on the grinder and place it in the water. Move on to the next piece, give it a pass on the grinder and place it in the water next to the first piece. Work the pieces in order and you will find the steel you pull from the bucket to grind has sat long enough to be cool to the touch. When you are

Fig. 90. *Steel with shards broken off.*

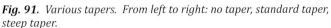

Fig. 91. *Various tapers. From left to right: no taper, standard taper, steep taper.*

Fig. 92. *Different thicknesses of steel.*

done, you will have five sections of well-ground and tempered spring-steel. Failure to cool your steel while grinding it will result in a wedge that has a soft edge. This will split, chip, bend, quickly dull, or fail during use. Well-made spring-steel wedges should last a lifetime with only minor maintenance, such as sharpening.

Having a large arsenal of wedges sharpened and ready for use is helpful when digging for Herkimer diamonds. A variety of lengths, thicknesses, and tapers is also recommended.

Keeping these tools free of heavy rust allows them to slide against rock and one another more easily. If used regularly, no oiling of steel should be required.

Lengths of steel between 10 and 50 centimeters are the most common, with lengths up to a meter sometimes being used. Lengths above 50 centimeters are usually not very helpful. Most spring-steel has a built-in curve. Some miners prefer steel with lots of curve. It does have some benefits, but one serious drawback is that it "springs" when pounded. Therefore, straight, or nearly straight, steel is often preferred. Most spring-steel is between half a centimeter and three

centimeters in thickness. Having a variety of thicknesses is also helpful.

Using spring-steel wedges is considered by some to be an art. Even after years of use, one can encounter new techniques that provide advantages in certain circumstances. Some of the basics of spring-steel use will be covered here.

First, you will need a method of pounding the steel into a preexisting crack in the rock. A series of crack hammers and sledgehammers between four and twenty pounds in head weight are usually used for this purpose. If working at a commercial site, you should find out if the owners place a weight restriction on the hammers used on the property. Most often the largest hammer you can use with a single hand, while holding the wedge in position with the other, is recommended for starting the steel. Short-handled 8- and 12-pound hammers often fill this role. Once the first steel is secure, it should be pounded to depth using the larger, longer-handled sledges. It is important to pound the steel a reasonable distance into the rock before adding a second steel to create a stack. Normally, this distance should be 20 to 40 centimeters. The deeper the steel is set the larger the potential

Fig. 93. *First steels in rock.*

Fig. 94. *Double stack.*

Fig. 95. *Triple stack.*

Fig. 96. *Major stack.*

Fig. 97. *Reversed steel, note point facing outward on bottom steel wedge.*

for breaking off a large slab of rock. If this is not done properly, the steel will merely shear off a shard of the rock face instead of splitting free a block of rock. The single spring-steel wedge may break the rock in the desired manner. However, this is usually not the case, and the second steel should be inserted on top of, or below, the first to begin creating your stack. Most often, getting in a second steel is easier when it is set below the first.

Because the Little Falls Dolostone is a layered sedimentary rock, most productive cracks run horizontally and the weakest bedding plane will run in this direction. Weathered cracks and stress fractures often run in other directions, but the primary grain to the rock is horizontal.

Once you are pounding the steel into the rock, the ring of the steel as it is hit will change noticeably. Loose steel makes a duller ring while secure steel makes a higher pitched ping each time it is hit. Miners refer to this as "singing." When the steel sings miners are happy. This will mean the steel is secure and working its way into the rock. Every swing of the sledge is now moving energy into your lift. If it is not singing, it is insecure and can pop out unexpectedly.

After two steels have been inserted to depth, a third, followed by a fourth, and so on, can be inserted. Spring-steel sliding against steel on its side(s) goes into the rock far more easily than steel sliding against rock alone.

Reversing steel is also a common practice if the crack being exploited is wide enough to allow for it. It involves placing the wedge into the opening backwards. It will not be able to be hammered further in at any point because the sharpened edge is facing out. This method has several purposes. First, when multiple wedges have been pounded in, it can become very hard to hammer new wedges in between them. A reversed wedge makes this easy to accomplish. The reversed wedge also allows the pressure to be exerted further under the rock layer, making it less likely that the stack will shatter the rock or shear off a thin section of it. It is more likely to lead to a larger section lifting and then breaking. Finally, blowouts are less common when reversing steels.

In the figures above, stacks using steel all set in the same direction are used in one stack while the one next to it demonstrates better technique showing the bottom steel in a

Fig. 98. *Reversed steel is more effective in top right stack. The pink arrow indicates a poor stack that is shearing out. The yellow arrow indicates an effective stack with the bottom steel reversed.*

Fig. 99. *The "Danger Zone".*

Fig. 100. *Section of table rock.*

Fig. 101. *Huge blocks being broken free.*

Fig. 102. *Shattered table rock.*

reversed position. The results show the standard stack shearing off a fragment of dolostone and the reversed stack looks like it is not doing anything. In actuality, the reversed stack is exerting pressure deeper under the layer and will eventually result in a block of rock being broken free, and not just a shard of rock. This may sound logical and easy, but it usually involves using a preexisting crack that is wide enough to slide the steel in backwards (usually an uncommon occurrence). Once a section of rock is beginning to lift an opportunity to reverse steel may present itself. If it does, one should take advantage of the situation by reversing as many stacks as needed until the largest section of rock possible is lifted.

Steel being inserted into a stack will initially tend to pop out. The pressure created here can cause the steel to shoot out like a rocket. It can slide sideways, but if you are hitting the wedge square, its preferred direction will be straight back out of the stack. Entire stacks can blow out, but they will usually not have the same force as single steels. Never stand in a danger zone where you can get hit by popping spring-steel. Recreational collectors watching you work will love to hang out in this area because it provides them an unobstructed view of your work. In actuality, this danger zone is where no one in their right mind would want to be standing. Tell them to move.

Sooner or later you will begin to hear the low cracking sounds that miners call "talking." This is your indication that the rock has begun to break, or rip, free from the surrounding layers. It is always a welcome sound because to get to this point has usually involved considerable effort. Many experienced

Fig. 103. *Stack in a pocket.*

Fig. 104. *Plate steel.*

miners will let their steel sit for a short period of time once they hear the rock talking. They do this so the crack will extend on its own in the desired direction, usually along the rock's bedding plane, and not have the crack change directions prematurely due to too much pressure, exerted too quickly and closer than they would like to their stack. Some even go to the extent of letting it sit for a longer period of time so the wedges have plenty of time to work. Steel set one day under high pressure can be left overnight and sometimes found to have created cracks in the interim.

Once the block you are working on is loose, the wedges will feel loose or pound in very easily. The lift is complete. Now the block may be levered out or toppled using long bars. If the block has shattered, you will have created other cracks to exploit using spring-steel, chisels, or other forms of wedges such as large wood splitting wedges.

Old, emptied pockets can sometimes be used as a means to get steel into the rock. In this case, the pocket can be filled with plate steel, reversed short wedges, sledgehammer heads, or other suitable material in order to establish a stack and use them as leverage. How the stack is created depends greatly on the shape and dimensions of the pocket. This is a tricky business, but can be very effective in certain circumstances when all else has failed.

Small, specialized wedges have their place in collecting Herkimer diamonds if they can stand up to the punishment. Small, thin spring-steel wedges can get into cracks that larger ones sometimes cannot. Once a crack is forced open they can be swapped out for larger wedges that can hold up to being hammered on with larger sledgehammers. Soft steel wedges are also an option. Softer steel that lacks temper will bend more easily, allowing them to follow the contours of a crack better than stiffer spring-steel. If these are bent in use, they can often be pounded straight again without snapping into pieces. They can also be found with steeper tapers and used to force open cracks in difficult rock that will not accept spring-steel.

These techniques work nicely for upper layers of rock or highly weathered table, but solid table presents other problems. First, not only is the table layer much harder, there are seldom any cracks into which steel can be inserted. When this is the case, one needs to cape a slot to substitute as a crack. This is done using what is called a capeing tool and a small 1-, 2-, or 4-pound crack hammer.

Capeing

Capeing is the process of chipping away the rock to create an artificial groove that serves as a crack for the steel. The term is well accepted in the Herkimer diamond community but does not appear in the modern literature. The term was one used by stone masons, but the term has fallen into disuse except by Herkimer diamond miners.

Capeing is a slow process and takes great patience in order to make an effective slot. The slot needs to be deep enough to create a crack in the rock in the desired direction and not just simply splay off a section of the rock face. Normally, this slot should be 10 to 30 centimeters deep in order to work correctly. Mentally, this is tough work. To tap away at a section of rock, turning it into dust is a long process. It can take several hours of steady work to complete. There are very few miners who are patient enough to use this method; however, those who do are able to break and move rock that seems impregnable to others.

Fig. 107. *Small crack hammers and capeing tools. The three capeing tools are, from left to right, a modern commercially produced tool, a historic capeing or pointing tool, and a homemade capeing tool made from spring steel.*

Fig. 108. *A caped slot with inserted steel.*

Finding capeing tools anywhere else other than at the mines themselves is unlikely. Making your own is slightly more complex than simply turning automobile leaf springs into wedges but not impossible. First, you must begin with spring-steel that is straight and consistent in thickness. Half inch thick stock is a good starting point. Next, the spring-steel stock needs to be cut into a tapered chisel-like form with a steel band saw or other method that will not heat the steel to the point of losing its temper. A tip needs to be cut so that it is nice and hard. Grinding tends to build up too much heat. Manually filing the final tip is a better approach. Sanding all edges is also important since, unlike spring-steel wedges, this tool will be held in your hand the entire time it is being used. The tip can also be flared out — made wider — by heating the steel and working it on an anvil. This has the benefit of helping to prevent the capeing tool from jamming in the slot as you work. If the tool is perfectly straight, this final step may be unnecessary. Maintain the tip as it dulls by filing, not grinding.

Making capeing tools using spring-steel that has not been tempered is also done. This will require that you have a source for the untempered spring-steel and a means by which to temper it properly. The benefit of this approach is that the steel is easier to shape and sharpen before it is hardened instead of after tempering.

Pocket Excavations

Pocket excavations are the fun part of field-collecting. The serious work has been done, and now you get to see, for the first time, what no other human has viewed. With this phase of collecting comes the tendency to rush. It isn't as simple as reaching in and pulling out the crystals and should be approached with care. Not only have you just entered into an opening that is millions of years old and has seen little in the way of disturbances during all that time, but you did so in a violent manner. Smashing your way in with huge sledgehammers and splitting the crystal's dolostone cocoon with spring-steel should be the extent of your heavy-handedness. You have also done other things inadvertently that are detrimental to crystals, such as exposed them to the air, likely a different temperature, and allowed light to reach them. The crystals may even have rolled out of the opening and landed on the ground. This is very tough treatment for a work of nature that has no business being anywhere else other than deep below ground.

From this point on, the extraction of specimens should be nothing short of surgical if your goal is to preserve the crystals in the best condition possible. Larger Herkimer diamond pockets have the potential to hold world-class specimens. It is frustrating to see great specimens turned into road gravel because of poor extraction techniques. Even the tiniest nicks or chatter marks can result from lightly sliding one crystal against another. Once the pocket has been opened, all attention should be focused on preserving its contents in an undamaged state.

Early season collecting includes the risk of thermal shock to the crystals; however, mid-to-late season pocket excavation is not as tricky because the ground should be about the same temperature as the air. When a crystal is removed from the cold confines of a pocket and is exposed to warm hands, warm air, or is set in the direct sunlight, it is at risk. The result may be haziness or a milky coloration that develops internally or it could result in actually breaking the crystal. Damage from thermal shock is not uncommon, and experienced early-season collectors are always prepared to minimize it. The most common practice when doing an early season pocket excavation is to have a bucket of cold water nearby in the shade to set the crystals in as they are removed from the pocket. Most experienced collectors can tell you stories about hearing or watching crystals crack while being excavated early in the season.

In most situations, the pocket excavation involves reaching into the opening and pulling out the crystals. Some may be in natural clusters or attached to a loose section of rock. Most will be lying loose in the base of the opening, often in clay. Some may be visible while others may not be in clear view. Others may be firmly attached to the walls of the pocket and require chiseling in order to remove them.

Remove all loose materials from within the pocket before attempting to recover pieces on the wall rock. When the loose material is a crystal or crystal on matrix, do not hit or even touch the wall with it. When materials are stacked together, avoid the temptation to rock pieces back and forth to widen the opening or loosen them from the other materials. Start from the top down and only pull out entirely freed specimens. If they are not freed, then widen the opening to gain better access. Small crystals of less than a millimeter in size are often common and easy to miss. There can be literally hundreds of these hiding in the pocket clay and in tiny spaces on the edges of the opening. Sweeping out the pocket with a small whisk broom can be helpful in recovering the smallest crystals and unseen crystal fragments. Sloppy work at this phase causes dings, abrasions or chatter marks on crystals and quickly turns fine specimens into merely average ones. It will also make all the hard work to find the pocket a waste of time and effort.

In situ specimens (those still in place) from the walls of the pocket are removed next. Chiseling a Herkimer diamond off the wall of the pocket is tricky if you wish to recover the crystal as a matrix specimen. First, get a good look at the interior to be sure you are not going to smash an even better specimen hidden behind the one you are attempting to remove.

Look for fracture lines or seams that can be used to split the rock. Use only as large a hammer and chisel as is required. Avoid overkill. If you can split the specimen free with a smaller tool like a screw driver then do so. A gloved hand filled with padding secured over the crystal pushing against the matrix while chiseling will increase your chance of getting the specimen off intact. This is where having a second set of hands to help is welcome. Packing the entire pocket with padding, such as newspaper, an old towel, or paper towels is also helpful in some cases. Removing the section of rock you are after with a single blow is going to save stress on the specimen. Knowing how much force is required to split it off the wall with one strike, yet not bash the specimen into oblivion, is a skill learned with experience. The goal is to avoid having the freed specimen fall into the base of the pocket or go flying off into another wall of the opening and, thereby, get damaged.

Taking a larger section of matrix than you would like your finished specimen to have is not a problem. Using a mechanical trimmer later can refine the specimen to the dimensions desired.

Keep all the contents of good pockets for closer evaluation at home. Sometimes a crystal fragment that is critical to the aesthetics of a specimen will be overlooked on site while at home, it could become important. Broken crystals, matrix, druse, pocket clays, and other materials can be laid in one or more cardboard flats while the good crystals or specimens can be wrapped in newspapers and kept with these materials. At home, reconstructions are easier if you have all the parts from within the pocket. Fragments and smaller crystals often hide themselves in the pocket clays and can be found by using a fine screened mesh and lots of running water.

Developing a system of organizing new finds can keep you from losing your sanity later at home. Things like wrapping accidentally broken crystal fragments together and in position saves trying to sort and identify those fragments later. Never mix contents of one pocket with those of another pocket. Rookie mistakes like this just double your work when reconstruction begins. Think of the pocket as a puzzle you just purchased at a garage sale in a plastic bag with no box. You don't know what the final picture looks like, so you must rely on how everything fits together to find success. A single missing piece will mar the final product, so making sure it is complete is critical. Mother Nature left nothing out; all the pieces are in the pocket, you just have to be sure you recover them. If you have everything from the pocket, a masterpiece can be reassembled.

Remember that when you find a pocket with crystals *in situ* on the pocket walls, there is no need to rush the process of specimen extraction. Think about what you are doing and visualize the approach that you intend to take. Get a more experienced miner's opinion or come back to the extraction after you have thought things through. Many fine specimens have been obliterated during careless extractions, so a little extra time put into doing it right may preserve that nice Herkimer diamond specimen.

Some pockets provide numerous obstacles to good pocket excavations. The one shown next is an example of this. The

Fig. 116. *Drilling (top) and blasting (bottom). Herkimer Diamond Mines Resort. Summer 2007. Jeff Fast photographs.*

Most true Herkimer diamond enthusiasts consider these methods of mining to be bad form or "cheating," so they will not be discussed further here.

Other Tips

When working high walls, the rock composing the upper layers normally gets removed first to reach the table below where the pockets are found. This broken waste rock can quickly become a boulder pile in front of your working rock face. You might just think that the rock can be broken down into manageable-sized chunks and rolled or thrown out of the way. This is sometimes the case, but all too often the pieces are so large they become unmanageable and accumulate to the point of interfering with your gaining access to the table layer. The result might be that although you have successfully worked your way down to the table, you end up having no

room to set steel or swing at it with sledges. Planning ahead for this is critical; otherwise, you may be doing all the work for nothing if you can't get at the pocket layer. Some of the mines will use heavy machinery, such as track hoes or excavators to help you remove this material, but it will cost you some serious money to have this done.

Preceding a single layer at a time when working a high wall is good advice. Going after multiple layers of overburden more than a meter thick can be risky and can result in a lot of wasted energy if the layer cannot be broken. Take on manageable sections while still making good progress through the rock. Layers of three meters or more in thickness can be broken free on occasion, but they still have to be dealt with once they are loose.

Work vertical fractures when they are available. Weathered seams can present themselves as vertical and diagonal openings in the rock that can be exploited. Taking advantage of these presents another means to gain access to pockets.

Working the ledge rock layer, found in some quarries below the table layer with the large dome pockets, is done in a similar manner as working the wall using spring-steel. About the only difference is that if someone has already harvested the pockets and left the ledge, there will not be much overburden to remove. Some individuals specialize in dome pockets, some in ledge rock, and others will mine either.

This is what real Herkimer diamond mining is all about. It is serious work. One should be in decent physical shape and prepared for the rigors of collecting in some of the toughest rock in the world. After several days of this type of work, no one ever wonders why the cost of the best Herkimer diamond specimens is so very high.

The repetitive motions associated with this type of work are tough on your body. Recovery time is important. Also, changing your tasks frequently can help. Pounding steel or capeing for extended periods of time is not recommended. Switch things up throughout the course of the day. Set some steel and pound it home. Leave it to lift the rock for an hour while working a vertical crack or emptying a pocket. Don't kneel all day working pockets or the dirt; occasionally stand and work some wall. Learn how to swing a sledgehammer both "lefty" and "righty" to balance the strain on your back. Roll or bar boulders instead of lifting them. Lift with your legs no matter how stiff you feel. Stretch out before, during, and after digging. Take care of your injuries. If you have been emptying pockets, you may have dozens of small cuts on your hands that could become infected. Wet down a napkin with rubbing alcohol and wipe them off and bandage the more serious cuts. Torn muscles need rest; they don't heal if you continue to work them. Keep your tetanus shots up to date. These are all simple things to do, but they often get lost in the excitement of searching for Herkimer diamonds.

Chapter 6:
Once There, Where to Dig

Once you arrive at a locality such as those discussed in Chapter 4, how do you figure out where to dig? This chapter provides information about how to find crystal-filled cavities in the rock once you are at the right place and have the right tools.

The region where Herkimer diamonds are found is huge. Hundreds of square miles hold the potential for finding these lovely quartz crystals and only a small percentage of the area has ever been excavated. The known locations represent a mere fraction of the potential places from which Herkimer diamonds could be recovered. Although much of the region has surface rock that is barren of pockets or rock that is covered by extensive glacial overburden or rock that is privately owned and off limits to collectors, other places have well exposed ledges or rocks with pocket levels near the surface or thin coverings of weathered soils above rock-holding crystals.

Commercial Locations

Watching collectors at the commercial collecting locations reveals several approaches to selecting actual spots to collect. Some collectors find a spot rather quickly and settle in to trying to uncover crystals right away. This approach may come from their knowing about where good material has previously been harvested, conversations with more experienced collectors or the mine owners, or simply a faith that they will have luck wherever they dig.

Others have a different approach. These individuals wander around, almost as if they are waiting for someone or something to tell them what to do. Sometimes, they will talk to collectors who are doing well, always a good approach when at a site unfamiliar to the collector. They may search for fresh diggings that show signs of someone else's successes. With Herkimer diamond collecting, this is often where they see large open pockets that have been emptied by previous collectors.

Going to areas where other collectors are working and having success must be done carefully and politely. Some will wait for the initial collectors to move on and then jump into their location while others will set up camp as close as possible. Generally, farther away is better than closer. Many collectors frown upon other collectors encroaching on their active digging site. Herkimer diamond digging is difficult enough without having to worry about others crowding your workspace and constituting a safety hazard. Flying rock chips and steel shards from tools are a real danger and should make the arriving collector respectful of an active site and careful about working too closely to others. Therefore, you should find a place to dig with some minimal distance from others around you—at least a great enough distance to provide some margin of safety should rock pieces or chisels inadvertently go flying. Asking a digger if it is okay to dig right next to them is rarely greeted with a denial, so it is smart and common courtesy to ask for this permission.

Claims, although informal, should always be respected. At some locations, they will be marked off by ropes, and sometimes they will be covered by tarps. Claims are usually verbal agreements between property or mine owners and collectors, and they often involve the exchange of money or the promise of a share of any crystals found. They are seldom legally binding, and so do not fit the standard formal definition of a claim, which involves the staking of boundaries and filing of documents with a government agency. At some commercial sites, informal claims are not allowed; however, a collector can always return to the same spot the following day. In these cases, collectors will often leave tools there or cover the spot with a tarp in order to "guard" their right to return and dig in the same spot the next day. If someone's tools are at a given spot, you should consider it spoken for and not dig there.

If you are a more seasoned collector, keep in mind that beginners often do not have a clue. They can get caught up in the excitement and will do or ask things that seem really naïve. When you open a pocket, you are bound to attract an audience. Having someone ask if they can take the diamonds out or chisel away a matrix specimen they see sitting there might seem to most logical people to be completely inappropriate, but it happens! People get caught up in it all and sometimes have to be told the obvious, "Sorry, I didn't do all this work so that I could give the crystals away, but you're welcome to watch from a safe distance." Inviting someone who is curious to sit in a safe location and watch you excavate the pocket is not a bad thing. It may be the only time they ever get to witness

Collecting Herkimer Diamonds from Loose Rock

Beginners are seldom prepared to tear down wall; however, they are often well prepared to break down moderately sized chunks of dolostone left behind by other collectors. Grapefruit- to basketball-sized pieces of dolostone can be broken with small crack hammers and chisels and will occasionally hold nice crystals within them. Larger blocks of rock can sometimes be found in the mining area. These can be broken up with the appropriate tools and also hold potential for finding crystals within them. Areas where more serious collectors are working wall become popular targets. In their rush to get down to the pocket layer, they frequently throw out sections of rock that may hold crystals. Care should be taken not to work directly behind miners using steel or throwing rocks into their tailing piles. Just ask where you can dig safely and not encroach on their work. More often than not, they will be helpful and may even point out the rocks that have a greater potential to hold crystals. If left to find these rocks on your own, it may pay to search the rocks left near to where previous good finds have been produced or rocks with a greater percentage of small pockets or vugs on their surfaces.

Perseverance

Luck may play a part in some important Herkimer diamond finds. There is always a story or two about someone finding a great specimen while simply walking along a trail or in a quarry after a rainstorm. Some stories are of children who stumble upon or break into large pockets. Collectors have been known to have great success on that first trip ever to look for Herkimer diamonds or on the one-day collecting trip they took with a group of students. Stories about beginners cracking into amazing pockets are not uncommon. Although wonderful happenings, these occurrences provide little useful information. Where the consistency lies is in consistency itself. Consistency, or better stated, perseverance, finds lots of Herkimer diamonds.

The best material comes from the rock and it is through the movement and breaking of that rock that great discoveries are made. Some might think it is oversimplified to think that the one who moves the most rock finds the best crystals but there may be a grain of truth in the idea. Being in the right area, using the most effective techniques, and staying physically strong all contribute greatly to success, but perseverance is far underrated as a primary factor that contributes to important finds. Once you have learned all you can about the methods, places to collect, and what signs to look for that indicate the presence of pockets, the playing field becomes level with everyone else who has this same advanced expertise. This is where perseverance takes over.

Collectors who work short days, collect only on weekends, avoid hard rock, and go inside during rainy weather, just do not move the same volume of rock as the dedicated collector with perseverance. Perseverance moves rock, and it is the moving of rock that exposes pockets. Which ones will hold the great crystals is impossible to determine. So, it makes logical sense to move lots of rock to increase your chances of success.

The dedication or commitment needed for digging Herkimer diamonds is not a trait acquired easily for some, and it is inborn with even fewer. The work is difficult in most respects. It is dirty and only rewarded occasionally when the results pay off with a find. Most of the time is spent doing strenuous labor and not pulling crystals from pockets. There is probably no upside to breathing rock dust or toiling for days at a time in the harsh sunlight. Those who can physically and mentally overcome these hurdles develop, or simply have, the perseverance required to get the job done. They will often seem obsessive-compulsive to others. It just doesn't make sense to the normal mind to undergo these types of stress willingly.

Even with all these, and other, negatives, there is no shortage of those willing to commit to the rigors of the work. It seems that the allure of finding Herkimer diamonds can be overwhelming and perseverance is rewarded. Every week during the central New York summers at various Herkimer diamond mines around the region, one can observe many serious Herkimer collectors hard at work.

I'm Not a Field Collector!

Individuals who love Herkimer diamonds, but want nothing to do with digging them out of the ground, need help too. Some guidance about purchasing Herkimer diamonds is appropriate. Various sources are out there and because these mineral specimens are readily available and popular there is an active market for them.

If you are visiting the source of the crystals, one of the commercial mines, you are likely to find Herkimer diamonds for sale. The mine's store or gift shop, local businesses, and other collectors are all potential sources. Even though most people have come to collect their own crystals, they may still supplement what they have collected by purchasing one or several specimens better or different from what was found. These better or different crystals may well be for sale in the site's gift shop. Don't assume they will be inexpensive just because you have found them at the source. Although mineral specimen prices can vary dramatically from place to place or dealer to dealer, the market for Herkimer diamonds is well established, and good ones are expensive everywhere. The miners who look like professionals and are collecting in the mines day after day are often selling their finds but will seldom advertise this fact. Talking with them can be a way to find good crystals for sale. Notice how hard they are working and don't go into shock when they tell you how much they want for their nicest finds. They normally will have established buyers already. Good Herkimer diamond specimens sell well and the very best ones are seldom even seen for sale.

***Fig. 119.** A nice selection of minerals at an area mineral show.*

Gem and mineral shows, especially the larger ones with a hundred or more dealers, will always have hundreds of Herkimer diamond specimens for you to choose from. There are even dealers who specialize only in Herkimer diamonds. This is where you are likely to find the best quality material. Dealers at the larger mineral shows get their inventory from a great variety of sources such as collectors selling their private collections, professional miners, and field collectors with excess finds. Some dealers will have only a single specimen while others will have hundreds. Comparison shop and you may find some good deals or run into the unique specimen that has the characteristics you desire.

Mineral clubs are a great way to enter the hobby of mineral collecting and to find people with a similar interest. Many also organize trips to the mines, and most will have collectors who have collected many fine Herkimer diamonds over the years. These individuals may wish to disperse some of their finds or at least let you see their collections. Often the best deals can be found with club members but it will take some time to meet the right person and may require refined negotiating skills.

Museum gift shops and local rock and mineral shops are a further source for Herkimer diamonds. Both these sources are likely to have lower-end specimens for sale, but some shops will have better pieces, as well. Always ask around if you are looking for really fine specimens. The shop owners may put you on a contact list and the day may come when they have that great specimen for you.

Nontraditional sources may have Herkimer diamonds for sale. It is not at all uncommon to find them offered at lawn and garage sales within the Herkimer diamond region. Also, they are often sold by the area's Amish population at their roadside vegetable stands. The source of these crystals is generally the plowing of their fields that are on Herkimer diamond-bearing horizons of bedrock.

The internet is an amazing source for Herkimer diamonds and allows the purchaser to access the world. You won't be able to handle the specimen immediately for close inspection, and that inability can be an issue for some people. Many full-time mineral dealers will have web sites for their mineral sales, and there will be thousands of specimens available for comparison shopping. These dealers almost always have "no questions asked" return policies so you may have the option to change your mind and get your money back. The eBay site is loaded with Herkimer diamonds for sale both by auction and by direct purchase. Sellers are evaluated by their customers and those evaluations are converted to a satisfaction rating for the seller.

The pricing of Herkimer diamonds has been driven by the demand for the specimens. Because of their beauty, the crystals of high quality and greater rarity have always commanded prices associated with other fine mineral specimens. This will often push quality Herkimer diamond specimens into the hundreds, sometimes thousands, of dollars in price. Generally speaking, clear, well-formed, euhedral crystals without damage command the highest prices. Crystals like this that are large, over four or five centimeters, are considered exceptional. Uncommon forms like scepters can bring even higher prices. Crystals on matrix or in aesthetic clusters also go for high prices, especially if they are naturals (not reassembled or reattached to the matrix). All Herkimer diamonds are unique in some way so the seller seldom has a hard time pointing out a multitude of characteristics that may make their specimen more valuable. Herkimer diamonds with damage, large contact marks from other crystals, those that are discolored or have distracting internal fracture or excessive inclusions, are avoided by experienced collectors.

Dealers selling Herkimer diamonds are noted for doing whatever it takes to make the specimen being offered for sale more attractive so it can command a premium price. This is often done by using terminology that makes the crystal sound rare when, in fact it is not. One example is the practice of calling a common characteristic, like having two crystals randomly connected to each other, something "special" like soulmates, penetration twins, or Siamese twins. (Note that actual Herkimer diamond quartz twins are exceedingly uncommon.) If the seller offers their product to the metaphysical market, there could be all manner of magical attributes given to these crystals.

When buying Herkimer diamonds from dealers, rock shops or online, it pays to negotiate a better price. There are thousands of specimens for sale and the asking price is just that. Many dealers will come down on their asking price if the request is politely made. The practice is more commonly accepted than the average consumer may be aware. The exception to this may be the truly exceptional specimen, which is rarely seen for sale. Dealers normally know that there are high-end mineral collectors willing to pay very high prices for the truly exceptional specimens. Very often, mineral specimens, including Herkimer diamonds, are highest in price near to the source. As a result, it pays to check out the inventories of dealers not within the Herkimer diamond region. They may not have as wide a selection of specimens but it is likely that their prices will be lower.

Another issue to keep in mind is that repairs, reassembly, and restorations, as mentioned in Chapter Seven, are common practices with Herkimer diamond specimens. Normally, if these modifications are well done, they will be impossible to detect. Some use ultraviolet light to detect adhesives but contrary to popular belief, not all adhesives fluoresce. The etiquette involved with reporting of these practices is not universally accepted. Some dealers assume everyone knows of these practices while others get their inventory from miners who do not inform them if the modifications have been employed. It is probably best to accept the fact that clusters of Herkimer diamonds have always been reattached to one another unless your information comes directly from a reliable dealer or collector who has first-hand information to refute that this common practice has been used on the specimen you are considering purchasing. So long as any repairs or reconstructions merely serve to return the specimen to its original appearance, downgrading such a specimen's price is unlikely. Dealers who highlight claims that Herkimer diamond crystal clusters are "natural", unrepaired, not reassembled, or not reconstructed are appealing to the fact that this is unusual in an effort to obtain higher prices.

Sometimes crystals that would not command a higher price due to damage will be modified and made to look free of damage. This is usually done by grinding the damaged area off and polishing the new surface or simply polishing the surface where damage is located. Polished faces often show slightly rounded contacts with other faces or do not show light striations or growth modifications seen on other faces of the same crystal. Knowing the crystallography of the Herkimer diamond is important in detecting modifications. Realizing the number of faces or angles between them is incorrect can detect these modifications. Magnification is also helpful as polishing swirls and lines may be detected. Unlike the failure to acknowledge that a specimen in reconstructed, failing to acknowledge that a specimen has been modified with lapidary procedures is dishonest. Most serious mineral collectors avoid such specimens.

Every knowledgeable collector should always be on the lookout for fakes. Herkimer diamond specimen fakes of great sophistication can be created because the high commercial value associated with them makes the practice attractive to some unethical sellers. The creation of a fake might involve such things as gluing together clusters of crystals that were not found together, attaching crystals to pieces of matrix that were not original, or creating fake enhydros or rainbows in crystals that did not have them naturally. A hand lens will not always find the sophisticated fake. Most of these methods are detectable with high magnification, preferably a microscope in the 30x range. With enhydros, ask for a guarantee that the enhydro will not "dry out." It might take several years but faked enhydros will usually dry out over time.

Detailed research makes it strikingly clear that the majority of the fine Herkimer diamond specimens have been collected in the past fifty years even though the crystals have been known for a much greater span of time. Their true value as top collector specimens has only been realized within this more modern time period. Most modern mineral collections and museums have fine examples.

Of the thousands of great examples of Herkimer diamonds collected in modern times, many have never been seen by the general public. They are not commonly offered by dealers at shows, shops, or online. When offered for sale, they are quietly dispersed to those who are well-known private collectors or managers of institutional collections that have expressed the desire to obtain the finest examples no matter what the cost. Some great specimens do eventually find their way into these elite collections and museums but the rest are rarely made available for public viewing. Many have been kept by those who have expended the Herculean effort to recover them from the ground. This does not necessarily mean they are squirreled away in boxes at locations close to the mines. People come from great distances to collect Herkimer diamonds and once addicted will come time and time again. These individuals, often young and not building great collections in the public eye, will hang on to their finest discoveries. In time, some amass important holdings that most others who appreciate Herkimer diamonds are unaware of, or if they are aware, are not privy to view.

As a result, there is a brethren of Herkimer diamond collectors, the true miners or field collectors, who know nothing of "silver picking" (collecting by buying). They are the ones in the forests and mines finding the remarkable specimens that other collectors only occasionally hear about and even more rarely see. They devote enormous amounts of time to searching out the elusive crystals and keep their most important finds as treasured keepsakes. They seldom show these treasures to others outside their circle of fellow miners. Often known only by their nicknames, these local legends are a secretive group and prefer to have their activities kept private and are most likely more successful at finding Herkimer diamonds because of this approach to their work.

Some within this group are older and are no longer active field collectors. Their crystals were found during the days of low walls when opening two or three major pockets a day was typical. Their collections often number in the tens of thousands of specimens and contain examples of some of the best Herkimer diamonds ever found.

Chapter 7:
Specimen Preparation

Occasionally, Herkimer diamonds come out of the ground in pristine condition, usually from pockets. They are clean, sparkling and have no stains, overgrowths, or excessive matrix. Seen like this, it becomes clear where the term "diamond" comes from. Most, however, need lots of help in order to become displayable mineral specimens. Not only will they often need to be cleaned, many will need to be trimmed and perhaps repaired in order to show their true beauty.

Trimming

For most collectors, the first step in specimen preparation is trimming the matrix to an optimal size. Obviously, this only applies to specimens that have their matrix preserved. Trimming can be accomplished with a simple hammer and chisel, but this will often cause crystals to be detached from the matrix during the process. To accomplish the desired end, screw-type or hydraulic trimmers can be employed. Both trimmers cause less stress to the specimen and are far less likely to cause damage to the piece. Further, they more accurately trim or split the specimen.

The screw-type trimmer is less expensive and easy to use. It simply brings two chisel points together using a hand-operated crank mechanism. Their limitation is that they can only apply a moderate amount of pressure to the matrix and are only capable of small trimming jobs. The hydraulic

Fig. 120. *Screw-type trimmer.*

Fig. 121. *Hydraulic trimmer.*

Fig. 122. *Nippers in use.*

Fig. 123. Top: *Matrix Herkimer diamond specimen before calcite removal.* Bottom: *After calcite removal.*

trimmers are stronger so they can trim down larger pieces while not inducing excessive shock to the specimen. They are far more expensive, but their versatility makes them a valuable asset to anyone who collects a large number of matrix specimens.

Sharp corners or edges on freshly trimmed specimens can be smoothed off with small nippers or an old pair of wire cutters. Very small specimens, referred to as thumbnail or miniature-sized specimens are also easily trimmed with nippers and, thereby, avoid being crushed by a larger trimming device.

Calcite Removal

Cleaning will usually come next, but prior to cleaning, there is a further consideration. Sometimes Herkimer diamonds form in conjunction with calcite. Usually this secondary mineral species is poorly formed, sometimes massive, weathered, or damaged. It can easily be removed with dilute hydrochloric acid. This product can be purchased from most hardware stores under the commercial name muriatic acid. It is 27 to 32 percent hydrochloric acid by volume. This should be diluted by adding one part acid to six parts water. The acid will attack the calcite quickly and the immersion should only take a few hours to complete. In some cases, additional acid will need to be added to the solution to keep the reaction sustained. As with the dilution of all acids, add the acid to the water, not the water to the acid.

Matrix specimens on dolostone represent a special case and require more careful consideration. The dolostone matrix will dissolve in the acid at a much slower rate than the calcite. Selective etching of the calcite alone can be best accomplished by using the dilute acid solution at low concentrations, lower temperatures, and for periods of time no longer than is necessary just to remove the calcite. Specimens can also be immersed in such a way as to allow the acid solution to touch only the areas coated by calcite. If you place a matrix specimen in an acid bath and forget about it overnight, you run the risk of ending up with a bunch of loose Herkimer diamonds and an ugly glob of etched dolostone.

After soaking in acid, the specimen needs to be soaked in fresh water for a period of time equal to its time immersed in the acid. Soaking the specimen at least twice as long as it was in the acid is preferable. This will remove any penetrating acid or residues of salts that could damage the crystals or the places in which they are stored. It may not seem like the acid would be able to permeate the dolostone or quartz but it does. This problem can

Fig. 124. *Completely cleaned specimen (Fig. 123). Crystal Grove mine, near Lassellsville, NY. 7.2 cm with crystals to 2 cm each. Collected May 23, 2011.*

Fig. 125. *Typical tools for basic mechanical cleaning.*

Fig. 126. *Solvent gun.*

be easily seen if the etched specimen is placed on cloth or paper for an extended period. The acid will leach out and cause damage to those materials, leave a residue on the crystals, or even stain the crystals slightly. This extended soaking or rinsing period is critical to the specimen's preparation and should never be neglected.

Hydrochloric acid is dangerous and needs to be handled carefully. Gloves and goggles are a must. Never use the acid indoors. The vapor is dangerous and will not only bother you, but it will eat away at metals. Use it outdoors, well away from children, pets, buildings, and vehicles, and always remember to cover it so the mist it creates while bubbling does not drift onto anything. The residue that remains can be neutralized with the addition of baking soda or any scrap calcite until the bubbling ends. The water and salt mixture that remains will kill plant life so it is best not to pour it on the lawn or near a favorite tree. It will effectively kill both.

If the specimen survives trimming and/or calcite removal, it may next require cleaning. Specimen cleaning is usually broken down into two categories—mechanical methods and chemical methods—both of which can be used to improve a Herkimer diamond's appearance.

Mechanical Cleaning

Mechanical methods can include very basic tools and techniques or highly advanced ones. Some of the simplest include scrubbing with abrasive powder, toothbrushes, wire brushes, 000 steel wool, dental tools or razor blades. The key to these mechanical methods is to be sure the medium that comes into contact with the Herkimer diamond is softer than quartz. The hardness of the quartz is seven on the Mohs hardness scale, so most abrasive techniques work fine on Herkimer diamonds. Mechanical methods usually involve scrubbing the material (e.g., steel wool against a stained surface) to remove the coatings.

More advanced techniques for cleaning Herkimer diamonds include the use of ultrasonic cleaning devices such as those used by jewelry stores. Micro-abrasive units like those employed for fossil development and preparation can also be helpful. Solvent guns used in the dry cleaning industry can be used with water for cleaning as can the dental appliances referred to as Waterpiks. Some people use Dremel-style tools with wire brush attachments to scrub the crystals. All these mechanical methods work well for removing pocket clays and some are good for removing heavy stains due to oxidation.

Chemical Cleaning

Yellow oxidation stains are common on Herkimer diamond crystals and are sometimes difficult to remove without using both mechanical and chemical methods. Chemical methods are more involved than mechanical ones, but can often be more effective. Chemicals can be used to reach areas between crystals or between a crystal and its matrix that would be impossible to reach otherwise. Also, chemical methods are less time consuming because they don't require continuous supervision as they work. The most common chemical methods for removing oxidation stains are the use of Super Iron Out, oxalic acid, or the Waller Method. All of these chemicals come in powdered form and must be dissolved in water. Specimens are then soaked in the solution for varying amounts of time.

Super Iron Out is likely the most commonly used method in North America for removing oxidation stains on mineral specimens of any type. It is readily available, inexpensive, does not produce toxic byproducts, poses less danger to human tissues than acids, and cleans up easily. Generally, it will not adversely affect Herkimer diamonds, associated minerals, or matrix if used properly. In the rare cases that associated calcite crystals have shiny faces, Super Iron Out may dull their luster. It is not readily available outside of North America. The powder is dissolved in warm water and the specimens are soaked for short periods of time; usually 20 to 30 minutes will do. Once submerged in the solution more powdered Super Iron Out can be sprinkled directly on the specimens. Super Iron Out can be used indoors; however, the fumes are an irritant to the respiratory system so containers with specimens left to soak should be covered. The fumes and the airborne powder should not be breathed in. Light-plastic food-storage containers work well for holding the solution.

Oxalic acid is more effective on difficult stains in tough to reach places but will require specimens to be immersed for much longer periods of time. This long term immersion also provides more opportunity for the acid to penetrate the cracks in crystals that may also hold stains. These cracks are often difficult, sometimes impossible, to clean effectively using any method. Follow the directions on the packaging carefully and remember to rinse or soak the specimens after their acid bath for an extended period of time. Oxalic acid is what most of the large commercial operations that mine quartz, such as those based in Arkansas, use to clean their specimens.

The Waller method is very effective but requires precise chemical ratios. Waller provides a detailed description of the process in *The Mineralogical Record*, 1980, and it is recommended the article be read in its entirety before pursuing the use of this method. A brief summary is described here.

The active ingredient in the Waller method is sodium dithionite. It can be difficult to obtain unless you live near a chemical supply house that carries this material. It can be difficult to get by mail, as well. The proper mix for the solution follows and should do a good job on most oxidation stains.

Sodium Citrate	71 grams
Sodium Bicarbonate	8.5 grams
Water	1 liter

This is your base solution. To this, add the sodium dithionite to create an active solution when you are ready to clean specimens.

Sodium Dithionite	20 grams

After soaking the stained specimens in this active solution, they can be removed, rinsed, and given a good day long follow-up soak in fresh water. Change the water several times over the soaking period to leach out any chemicals that have permeated the crystals, matrix or fractures within them.

It cannot be overemphasized that after any of these chemical cleaning methods is complete, the specimens should be soaked in fresh water to remove any remaining chemicals. A final rinse in distilled water will prevent water spotting.

The series of steps involved in the cleaning process of a typical Herkimer diamond matrix specimen may include the following:

1) Trimming off excess matrix to create an acceptably sized specimen;
2) Removing any distracting coatings of minerals such as calcite;
3) Mechanical cleaning with a toothbrush and soap to remove any remaining clays;
4) Immersion in Super Iron Out solution (first soaking) to remove brown iron oxide stains;
5) Removal from Super Iron Out solution followed by mechanical cleaning with steel wool;
6) Super Iron Out solution immersion (second soaking and more soakings if required);
7) Removal from Super Iron Out solution followed by soaking in fresh water for at least a day; and
8) Removal from water, rinsing in distilled water and air drying.

The transformation can be dramatic.

Fig. 127. *Cleaning steps 1-2. Step 1. Toothbrush scrub with soap and water to remove clays. Step 2. Steel wool and razor work to remove heavy oxidation.*

Fig. 128. *Cleaning steps 3-4. Step 3. First soak in Super Iron Out. Step 4. Removal from soaking and steel wool work on the few remaining oxidation stains.*

Fig. 129. *Cleaning steps 5-8. Step 5. Second soaking in Super Iron Out. Step 6. Removal from soaking in Super Iron Out. Step 7. Two day soaking in fresh water. Step 8. Rinse in distilled water and dry.*

Repair, Reassembly, Restoration

Sometimes broken crystals can be repaired or reattached to their original positions on matrix. This will require the use of waterproof glues that dry clear and hold firmly.

Two part epoxies are commonly used for this purpose, but keep in mind that they will discolor over relatively short periods of time. Many products are available with drying times that range from minutes to hours. Will you have to hold the pieces being glued together in place or will they stay in place if you just set them aside? This can be an important consideration when making a repair or doing a reassembly. If you are going to have to hold them in position, a fast-drying product will be preferred. Most of these products have a high viscosity and are thick and paste-like. During reassembly, glue often squirts out from between the crystals if too much is used. A little glue will go a long way. Fast-drying adhesives are certainly preferable for matrix reattachments or matrix repairs.

Cyanoacrylate-based glues, commonly called superglues, work well but have a tendency to fail over extended periods of time. Also available in many forms, they usually have lower viscosities that are often preferable when reconstructing Herkimer diamond clusters or repairing broken individual crystals. Hardening times vary from seconds to hours so fast curing products have an advantage when doing reassemblies

of clusters. Most of these adhesives use water as a catalyst so their curing time can be accelerated by leaving one of the surfaces damp. Superglues stick to almost everything including human tissues and so should be used with care.

UV-curing epoxy resins are preferred by some due to their ease of use. These epoxies only require that the crystals be held together when exposed to UV light (sunlight works well), and they lock into place without having to wait an extended period of time for the glue to set.

Curatorial solutions exist that have stability periods much longer than the aforementioned adhesives. These acryloids (ethyl methacrylate copolymers) such as B-72 will not yellow and have a shelf-life approaching 300 years according to the manufacturer. Acryloids come in pellet form and are dissolved into solution with acetone or toluene. Like all the other products, B-72 is totally removable (reversible) in solvents such as acetone. It is best used, however, in thinner solutions, as a stabilizing agent on crystal specimen matrices if the piece requires stabilization.

Normally, very small amounts of any adhesive are required to get the job done. If used in excess, glue will flow onto the surface of the crystal from the repair or reattachment and leave an unsightly residue. This residue can be removed with a razor blade after it hardens. Almost all of these adhesives dissolve in acetone, so it can be used for cleanup or to remove excess glue from surfaces of the specimen, as well. Keep in mind that repaired specimens are frowned upon by many in the mineral collecting world so any such maintenance to crystals should be noted on specimen labels if the specimens are to be sold.

The process of using glues to reassemble, repair, or restore crystals and clusters is reversible. This means the work can be disassembled and redone if the original job was substandard, the glue has failed or yellowed, or if crystals have been glued into the wrong positions accidentally or intentionally. The most common method of disassembly is to soak the specimen in acetone to dissolve the adhesive.

Clusters of Herkimer diamonds are almost always found with individual crystals loose from one another so most collectors will restore them to their original positions in the clusters by gluing them back together. This reassembly is an acceptable practice done with almost all Herkimer diamond clusters. Rarely are clusters found in groupings that have not come apart. Intact clusters are referred to as "naturals." Clusters of Herkimer diamonds should be assumed to be reassembled unless otherwise marked. This is the reverse of other mineral specimens, where a reassembled cluster of crystals not marked as such is considered unethical.

The process of reassembling clusters can be complex with groups that include tens of crystals. Matching of male and female separation surfaces can be time consuming even though it would appear simple. No two male and female contact surfaces are alike and a perfect joint can only properly be made between the two that match. The contact surfaces resemble fingerprints and when properly aligned fit together only one way. The following photos show two Herkimer diamonds that match for the simplest of reassemblies — two attached crystals.

Restoration involves returning a mineral specimen back to the form it is believed to once have had. For example, a centimeter corner of an otherwise outstanding 10-centimeter crystal is missing. The collector may elect to recreate the corner using an acrylic resin and mold this artificial corner onto the crystal. Another example would be that a nice crystal has a blowout on a prominent face, and it is filled using a resin. The use of restoration with Herkimer diamond specimens is rather uncommon. Most restoration materials will lack the internal characteristics of the quartz and will likely have a different refractive index that will be noticeable. Almost all people involved in specimen mineralogy frown on restorations.

Fig. 130. *Top to bottom, the reassembly of a two-crystal cluster.*

Chapter 8:
Mineral Collections

Collecting specimens in the field and acquiring them from others are only part of the process of building and maintaining a mineral collection. Documentation, organization, and safe storage of mineral specimens preserves their value as historical objects, scientific objects, and investments. Studying, displaying, and exhibiting specimens in a collection can greatly increase the enjoyment of the collecting experience. Although experienced collectors will have realized the importance of properly caring for their collection at some point, beginning collectors should get acquainted with some of these ideas as early as possible.

Minerals are interesting objects, especially when they are found as crystals. They pique the imagination on many different levels. These natural creations are attractive to young and old. Even for people with only limited initial interest in the natural world, seeing a sparkling crystal is often the beginning of an appreciation and is commonly the way beginners develop their interest in minerals. Some may narrow their direction, perhaps focusing on metaphysical issues or displaying specimens as home décor objects or working with their children to develop an interest in science, but some will end up collecting minerals as a hobby.

Mineral collections can take on many themes or none at all. Some collectors are generalists who enjoy any mineral and so collect whatever interests them at the time. They are likely to field-collect some of their own specimens, but are equally likely to supplement these with ones they purchase or acquire by exchange with other collectors. As they collect, they may elect to focus on a given aspect, or theme, or enjoy all aspects of mineral collecting and not specialize. However, those who become more serious will often develop a theme to their collecting depending largely on their personal interests. It is very common to build a collection of specimens from a particular geographic area, for example. If the individual does a great deal of their own field collecting, it is also likely to center around the area or areas where they do that field-collecting. Minerals from the Herkimer region or Herkimer diamonds in particular have been the theme of numerous important collections.

The first endeavor of the organized collector will be proper documentation of the specimens in the collection.

Such documentation might initially seem unnecessary for a collection of Herkimer diamonds since what they are is obvious and everyone knows where they come from. Actually, because they come from a region and not a single locality, documentation of the exact locality is essential to preserving its value as anything more than a pretty crystal. In order for that important information to outlive the collector and be available to future generations, documentation is essential. Obviously, this may not apply to every run-of-the-mill crystal that enters one's collection but it is certainly important for fine specimens that have the potential to be passed on to other private and institutional collections, or are unusual in some historical or scientific sense.

Most collectors create a catalog of their collection and provide the following data on catalog cards: *Identification Number*, which is applied to the specimen, catalog card, and label(s); *Location*, often very specific including mine, GPS coordinates, pocket level, etc.; *Description*, including size, form, damage, repairs, etc.; *Date Acquired*, when self-collected or when purchased and from whom or when received in exchange and from whom; *Date Collected*, if purchased or acquired by exchange the original date the specimen was collected; and *Collector*, who field-collected the specimen if it was not self-collected.

Purchase price, estimated value at the time, appraisals, old tags, labels, and other details are often helpful to the collector or later owners, as well.

A label that accompanies the specimen wherever it is stored or displayed is also recommended. These basics can be replaced by a digital archive that has the specimen's photograph(s), avoiding defacing the crystal with a catalog number; however, a label to accompany the specimen is still recommended. Digital documentation also allows for convenient updating of information and off-site storage of copies. Several commercial software programs are available to collectors for documenting their collections.

If the collection is rich in specimens collected by others, the history or provenance of each specimen becomes particularly important. Documentation of who collected the specimen and when, as well as the history of ownership up to the present, along with the original labels of earlier collector-owners often adds significant historical and financial value to the specimen. For these kinds of collections where the specimen is accompanied by earlier labels, special care needs to be taken to avoid separating the specimen from its labels.

When fine specimens are displayed in the home, loaned to institutions for research or display, cleaned, photographed, appraised or any other number of unusual events takes place, it becomes possible for the specimen to be separated from its label. Therefore, a catalog number or photographic record is invaluable. Reference can be made to the catalog card or image and the complete data are once again available. A new label can be generated if required. Specimens without detailed data become little more than pretty rocks.

The display of Herkimer diamonds is fairly straightforward with the exception of thermal control. Herkimer diamonds should never be frozen or allowed to remain in direct sunlight for any reason. Even though they may not be visible to the eye, fluid-filled cavities are common in Herkimer diamonds. To allow them to freeze, or worse yet, heat up in open sunlight, can provide opportunity for the water inside them to expand. This might just result in a small blowout marring the surface of one face, but can result in the complete destruction of the crystal. Because water expands both as it freezes and when it vaporizes, Herkimer diamond specimens should be stored in a controlled thermal environment that avoids both.

Storage of Herkimer diamonds can be accomplished using cases with shelves, or better yet, closed drawers. Closed drawers keep out light and dust and can help the collector organize the collection. Since Herkimer diamonds rarely attain large sizes, small drawers work very well. Even large clusters generally tend to be two-dimensional in nature and can be housed in drawers that are relatively shallow. Shallow drawers mean more drawers taking up less floor space — a plus as one's collection grows.

When displaying Herkimer diamonds in a case or curio cabinet that uses a lighting system, knowledge about any heating caused by the lights is essential. Incandescent bulbs, even small, low-voltage lights, are notorious for heating the inside of display cases to unacceptable levels. The heat in the case will rise usually making the uppermost shelves more vulnerable to thermal shock. Direct venting with small fans or passive venting by having holes in the ceiling of the case may solve the problem; however, other forms of lighting might be safer for the specimens. The trend for mineral display seems to be moving toward halogen spotlights and more recently to LED (light emitting diode) lights that come in strips that can be mounted along the vertical or horizontal edges of the case. Although most Herkimer diamonds lack significant color, accessory minerals and specimens of other minerals will appear with best color accuracy if the lights have a color temperature of 4,500 to 5,000°K.

Dozens of varieties of display stands are available for mineral specimens. These allow positioning of the specimen at an optimal viewing angle. Inexpensive standard sized bases can be purchased at mineral shows or online. Custom bases can be made as well. Adhesives that temporarily bond the crystal to the base are available. Normally these putties provide stability and do not dry out quickly. The colors, however, are sometimes distracting. Some collectors use hot glue effectively for the same purpose. Custom-made acrylic bases usually require no adhesive for specimens to remain in stable positions. Some collectors avoid stands of any type finding them distracting and unnatural.

Small display/storage boxes are often used to house individual specimens of smaller sizes. Commonly called perky boxes, they provide a way to organize individual crystals while keeping critical data with the specimen.

Housing or displaying the sulfides that are associated with Herkimer diamonds, especially pyrite and marcasite, can be problematic due to a phenomenon variously referred to as pyrite disease, pyrite rot or pyrite decay. Pyrite disease results from the aqueous oxidation of iron and sulfur to release ferrous iron and sulfate ions into solution. Since the oxidation occurs from aqueous solution and not directly from the oxygen in the air, keeping specimens as dry as possible is the best approach. The problem is most common in iron sulfide minerals from sedimentary environments largely because the granular, layered, or sheet structure of pyrite and marcasite specimens formed in these environments increases the surface-to-volume ratio of the material. The breakdown is irreversible but can be prevented or retarded with limited success by following these steps. 1) Inspect your collection often. It will be obvious if you have a problem. The sulfide will dull, white efflorescences will appear on the surface, and sulfuric and sulfurous acids will degrade whatever the specimen sets on. If not checked, pyrite disease will ultimately cause the specimen to crumble. 2) Control the relative humidity, keeping it as low as possible, preferably 35% or less. 3) If you know you have a problem, act quickly and clean the powder off the specimen and take efforts to house it in lower humidity conditions. Isolation and desiccants can help as can washing thoroughly in a solution of sodium bicarbonate to neutralize the acids produced that speed up the oxidation reactions. However, particularly thorough drying is then required. Do not let your labels or calcite specimens come in contact with these diseased specimens. Calcite will react with the byproduct of this reaction, sulfuric acid, and should not be stored together with sulfides if possible.

Chapter 9:
Variations on a Theme

Herkimer diamond specimens come in so many shapes, sizes, crystal forms, and combinations that it almost defies imagination. The main types are described here.

Single Crystals

Single crystals can reach large sizes up to several kilograms, but more commonly, they are small, measuring one or two centimeters in length. They can also be so small that they look like sand.

Smaller, transparent, damage-free crystals are often referred to as jewelry grade crystals. Some individuals, especially mineral dealers and jewelry makers, use a grading system to simplify their descriptions. The codes AAA, AA, A, B, and C are common in commerce, but have definitions that vary somewhat. The clearest, flaw- and inclusion-free crystals get the higher grades (AAA or AA), while crystals with imperfections receive lesser grades. Almost all Herkimer diamonds of several centimeters or larger will show some internal imperfections. Crystals exhibiting no external damage or contact marks from other crystals or the pocket matrix are universally preferred and commonly called floaters due to their obvious lack of attachment to anything. Unfortunately, there is no accepted standard when it comes to this form of grading and no governing body, such as the Gemological Institute of America (G.I.A.), that monitors grading or does appraisals using a universally accepted method.

Fig. 131. *Extremes in size: A large single crystal weighing two kilograms beside 690 small crystals in a teaspoon. It would take more than 1,000,000 of the small crystals to have a mass equal to the single large crystal shown here.*

Fig. 132. *A grouping of jewelry grade single crystals. Ace of Diamonds mine, Middleville, NY. One half cm to 3.5 cm each.*

Clusters

Many crystals form together in interconnected groups called clusters. When found, they are usually not together in their original positions, but have come apart and are found loose on the floor of the pocket. The cluster may have once been attached to the pocket's interior or could have formed with no attachments and be called a floater just like single crystals with this characteristic. Collectors tend to reassemble, or reconstruct, these clusters into their original arrangements using adhesives. Almost all Herkimer diamond clusters have to be reassembled in this manner and the practice is well accepted as normal. Loose Herkimer diamonds that were once attached to other crystals in a cluster will have unique marks on them that match the mirror image mark on the crystal to which it was connected. Reassemblies can often be rather involved processes that take a surprisingly large amount of time. Broken crystals are often glued back together, but this process is called a repair. Repairs to crystals are not usually viewed in the same positive light as the reassembly of clusters of crystals. Clusters that have not needed to be reassembled are rare and referred to as natural.

Druse

The dolostone on which Herkimer diamonds often form can be barren, or in some cases, coated by tiny crystals of quartz referred to as druse. This druse can be composed of colorless crystals or those made dark brown or black by included anthraxolite. Specimens with colorless crystal druses tend to be rather common, whereas those included by hydrocarbons are far rarer, more aesthetic, and significantly

Fig. 134. *Black druse specimen. Herkimer Diamond Mines Resort. Field of view, approximately 5 cm. Collected by Eric, John, Brett and Jay Medici and John Hinkel, 1986. Medici collection. John Medici photograph.*

Fig. 133. *Herkimer diamond cluster.   Ace of Diamonds mine, Middleville, NY. 7 cm. Collected by Tom Dillon.*

more valuable. They are often called black druse specimens and they represent what some would consider to be the ultimate in collectable Herkimer diamond specimens.

Matrix Specimens

Herkimer diamond crystals sometimes form directly on the dolomite rock in which they grow. When this is the case, the specimen is referred to as being on matrix. When the crystal and matrix are removed together, the resulting specimen is called a matrix specimen. Sometimes matrix specimens are found loose in the pocket; in other cases, they have to be chiseled free in order to be removed. Most collectors prefer matrix specimens so an effort is almost always made to keep the crystal(s) attached while removing the rock. If crystals have broken off naturally or during removal, they can be reattached to the matrix in their original positions. Again, this is not preferable to unrepaired, natural crystals on matrix. Terms are sometimes blended: matrix clusters

Fig. 135. *Herkimer diamond on matrix. Crystal Grove mine, near Lassellsville, NY. 6 cm specimen with 3.3 cm crystal. Collected June 3, 2011. Jay Walter collection.*

Fig. 136. *Parallel growth crystals. Treasure Mountain mine, Fall Hill, Little Falls, NY. 5.1 cm. Collected Summer 2002.*

or singles on matrix are commonly used to describe specimens. When detached crystals are glued back into the original position on the matrix, they would be called reattached, which is a form of repair. On the other hand, if crystals are glued to a matrix on which they did not form they are called fakes or counterfeits.

Parallel Growth

Parallel growth is when two or more connected Herkimer diamond crystals have identical crystallographic orientation and thus have faces that are parallel to one another. It is most noticeable when the specimen is turned in the light and the reflections off multiple faces are noticed all at once. Parallel growth is a relatively common characteristic in Herkimer diamonds.

Twins

Twinning in quartz is a complex issue and not as obvious as some would imply. Many references to twinned Herkimer diamonds are mentioned in print and online. Most noted examples are actually either crystals that simply interpenetrate one another according to no definable twin law or are crystals that are in parallel growth.

This does not mean that twinned Herkimer diamonds do not exist, however. Dauphiné, Brazil, and a combination of the two

twin-laws were noted by Gault (1949) by observing the external twin boundaries on the surfaces of crystals, which he revealed by etching the crystal's surfaces with a corrosive solution such as hydrofluoric acid. He found the phenomenon to be quite common with the combination of Dauphiné and Brazil laws to be almost twice as frequent as either form alone. Parallel growth Dauphiné and Brazil twinning and a combination of the two do exist in Herkimer diamonds; however, failing such etching, this kind of twinning is unidentifiable just from looking at a crystal's external morphology. The existence of actual Dauphiné or Brazil law twins that can be identified in hand specimens is exceedingly rare because the secondary crystal faces that show such twinning are usually not present on Herkimer diamond crystals (see Appendix C). It could likely be one of the rarest of all variations in Herkimer diamond specimens.

In the case of Brazil-law twinning, the twin plane is perpendicular to an "a" axis of the crystal. A pair of "x" or "s" faces should be visible clearly demonstrating that both left and right-handed quartz crystals are present indicating the existence of the penetration twin. The Dauphiné twin will show itself through twinning along the "c" axis and will manifest itself with "x" or "s" faces showing the presence of two left-hand or two right-hand crystals. In these interpenetrant twins, the composition surface is not a plane and is not visible to the observer. By contrast, Japan-law twins form as mirror-image

Fig. 137. *Dauphiné twin, exhibiting two elongated parallelograms that are "s" faces (trigonal bipyramids). Treasure Mountain mine, Fall Hill, Little Falls, NY. 1.2 cm. The "s" faces have been darkened for emphasis.*

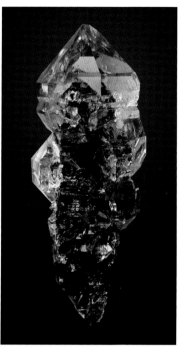

Fig. 138. *Classic scepter. Treasure Mountain mine, Little Falls, NY. 6.5 cm. Previously in the Robert Whitmore collection.*

Fig. 139. *Skeletal scepter. Treasure Mountain mine, Little Falls, NY. 6.5 cm.*

crystals on either side of a composition plane. No documented examples of Japan-law twins in Herkimer diamonds exist.

Scepters

Many minerals, including quartz, occasionally show a peculiar growth habit consisting of an elongated crystal with a bulge at one or both ends, resembling a royal scepter. These crystals are fittingly referred to as scepters. In the case of Herkimer diamonds, the stem of the scepter is almost always impregnated with black anthraxolite and the bulged overgrowth at the tip is colorless. A detailed labeled drawing appears in Appendix D. These complex and uncommon crystals develop first as an anthraxolite-included prismatic crystal. After this initial development, selective growth of more, usually colorless, quartz occurs on the termination of the crystal. This colorless overgrowth will be in the same crystallographic orientation as the black prismatic stem. True scepters always show this parallel orientation; however, some quartz scepters from other types of localities do not and are thus mislabeled. In other words, a quartz crystal growing on top of another quartz crystal is not a scepter if it lacks this crystallographic alignment.

If this later growth occurs on both ends of the crystal (both terminations), the crystal is referred to as a dumbbell or barbell, due to its resembling the dumbbell one would

find in a weight room. In rare cases, they form in groups or have branches off the main crystal. They often show complete overgrowth by the clear late stage quartz and are then called internal-shaft scepters. In this case, the crystal's black prismatic shaft would be completely enclosed giving the specimen the appearance of a morphologically typical Herkimer diamond with a large prismatic black crystal inclusion. Locals who specialize in mining or collecting scepter Herkimer diamonds tend to call all crystals with any form of prismatic, anthraxolite-included stem and some late-stage overgrowth, a scepter, even if they do not exhibit the classic form. When the morphology is very complex, not showing the classic scepter form, they sometimes refer to these crystals or groups of crystals by the inelegant term "gobstoppers". The variations of these crystals are almost endless. Scepters with skeletal growth on the enlarged tip, smoky tips, flattened tips, tips showing rainbow inclusions, and clusters all exist. So-called rider crystals found attached to the surfaces of scepters are always found in parallel growth with the scepter stem and thus multiple riders form a parallel growth. The only exception to this parallel alignment of different phases of growth in scepters appears to be when they have branches growing off the main body of the crystal. These branches typically exit the stem of a scepter crystal with no particular crystallographic orientation to the main crystal. This may imply that branches

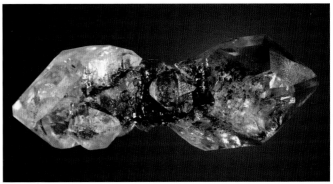

Fig. 141. *Dumbbell scepter. Treasure Mt. mine, Little Falls, NY. 6.7 cm. Brian Norensky specimen and photo.*

Fig. 140. *Branching dumbbell scepter, Treasure Mountain mine, Little Falls, NY. 6.6 cm. Steve Chamberlain collection #15575 and photograph.*

Fig. 142. *Scepter cluster. Treasure Mountain mine, Little Falls, NY. 9 cm. Steve Chamberlain collection, #15271. Collected by the author, August 24, 2001.*

are actually separate crystals that became connected to the main crystal during growth.

Reverse scepters also have an anthraxolite-included stem but no overgrowth on the tip. The name is something of a misnomer. Instead, the initial prismatic crystal tapers down to a smaller-diameter termination. Both scepters and reverse scepters can be found in the same pocket or occasionally in the same cluster of interlocking crystals. The exact mechanisms behind the differential growth that leads to the formation of both scepters

and reverse scepters is not fully understood and is debated among Herkimer diamond experts and specimen mineralogists.

Tabular Crystals

Crystals that are tabular in habit, often called "tabbies," are not uncommon. They can form as isolated single crystals or in clusters with traditionally shaped crystals. They look thinner perpendicular to the crystal's length and appear somewhat blocky in form.

Fig. 143. *Reverse scepter. Treasure Mountain mine, Little Falls, NY. 5.3 cm. Collected August 24, 2001. Jay Walter collection.*

Fig. 144. *Reverse scepter. Treasure Mountain mine, Little Falls, NY. 3.4 cm. Collected summer, 2001.*

Fig. 145. *Tabular crystal on matrix. Crystal Grove mine, near Lassellsville, NY. 3 cm. Collected May 30, 2011.*

Skeletal Crystals

Skeletal crystals, more accurately called Fensterquartz (German for window quartz), are an interesting variation that occurs when the edges of crystal faces grow at the expense of the centers. The end result looks like a highly fractured or broken crystal. In extreme examples, the body of the crystal appears to be made of thin, parallel plates of razor-sharp quartz. Skeletal crystals normally form as single, isolated crystals instead of clusters at most locations and are often found in pockets with traditional Herkimer diamonds. Large skeletal crystals are common at some locations like those in Middleville, while other locations like those on Fall Hill, are dominated by smaller ones.

Sometimes, later in their development, skeletal crystals get covered by further stages of quartz that results in the skeletal faces being encased within a traditionally-shaped crystal. Because the skeletal faces have large gaps between these quartz surfaces, a convenient place is created for materials to be trapped as inclusions during the crystal's growth. These inclusions are another important aspect of Herkimer diamonds and, at most locations, tend to be gases and solids, often pocket clays and hydrocarbons, and occasionally liquids.

Fig. 146. *Slightly smoky skeletal crystal. Treasure Mountain mine, Little Falls, NY. 6.8 cm. Collected Summer 2000.*

Fig. 147. *Skeletal crystal cluster. Ace of Diamonds mine, Middleville, NY. 19 cm with crystals to 11 cm each. Jay Walter collection.*

Fig. 148. *Prismatic crystals with phantoms, calcite, dolomite, pyrite, and hydrocarbons. Lowville, Lewis County, NY. 6.1 cm specimen with largest crystal measuring 5.5 cm.*

Prismatic Crystals

Prismatic crystals, crystals elongated along the "c" axis, are what typical quartz crystals look like. Although not uncommon in the Herkimer diamond region and the surrounding areas that produce similar Herkimer-style crystals, they are not the norm. It is unusual for the elongation to be excessive. A Herkimer diamond crystal that is more than four times longer than it is wide is exceedingly rare.

Inclusions

Even more diverse than the Herkimer diamond's external variations are the variety of things found within them. Collectively called inclusions, these internal imperfections add diversity to the Herkimer diamond's appearance. Many collectors feel they tend to distract from the perfection one associates with these crystals. From a more scientific point of view, these inclusions provide clues to the origins of the Herkimer diamonds themselves.

For almost 200 years, scientists have studied inclusions in Herkimer diamonds to try to better understand how the crystals themselves formed. The topic is beyond casual reading for most but is well summarized in a paper by Tuttle, 1973. Early work was conducted on fluid inclusions to determine the solution's composition by Sir Humphrey Davy (1822). More modern works (Keith and Tuttle, 1952 and Dunn and Fisher, 1954) on solid inclusions better describe the composition and properties of the material called anthraxolite (see Appendix E). Dunn and Fisher use their work to go further and provide a sequence of events for the formation of materials present in many Herkimer diamond deposits (see Appendix F).

Types of Inclusions

Inclusions, usually grouped into three categories — solids, liquids, and gases — can be trapped within the Herkimer diamond's structure. Inclusions are usually small but examples to a centimeter or more are known.

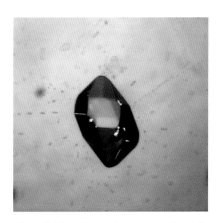

Fig. 162. *Negative crystal with surrounding, oriented micro-inclusions, some in stringers. Treasure Mountain mine, Fall Hill, Little Falls, NY. Collected 2002.*

Fig. 163. *Negative crystal with solid, liquid, and gas inclusion (three-phase). Dendritic hydrocarbon in movable gas bubble. Treasure Mountain mine, Fall Hill Little Falls, NY. Collected 2002.*

Fig. 164. *Black surrounded by brown hydrocarbon inclusions. Lowville, NY.*

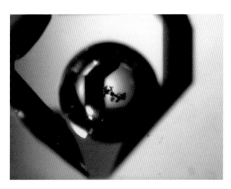

Fig. 165. *Negative crystal with solid, liquid, and gas inclusion (three-phase). Dendritic hydrocarbon in movable gas bubble (close-up of Fig. 172). Treasure Mountain mine, Fall Hill, Little Falls, NY. Collected 2002.*

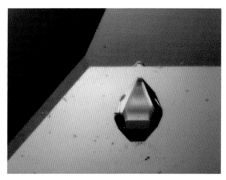

Fig. 166. *Brown hydrocarbon in negative crystal. Lowville, NY.*

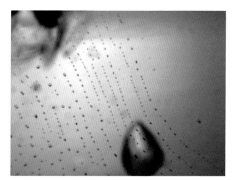

Fig. 167. *Oriented micro-inclusion stringers in parallel formation. Treasure Mountain mine, Fall Hill, Little Falls, NY. Collected 2002.*

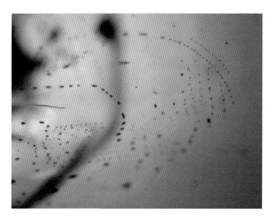

Fig. 168. *Oriented micro-inclusion stringers in parallel formation. Treasure Mountain mine, Fall Hill, Little Falls, NY. Collected 2002.*

Fig. 169. *Elongated negative crystal with movable gas bubble in liquid and brown hydrocarbon (three-phase). Lowville, NY.*

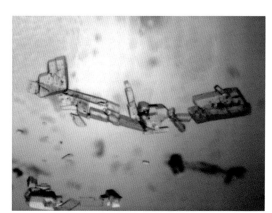

Fig. 170. *Rhombohedral shaped inclusions of unknown material. Treasure Mountain mine, Fall Hill, Little Falls, NY. Collected 2002.*

Fig. 171. *Partially formed negative crystals. Lowville, NY.*

Rainbows

Rainbows can sometimes be seen in Herkimer diamonds and result from the same thin-film optical effects as when oil coats water. They can result from petroleum or water that has penetrated along cracks or thin fractures in the crystal. The layer of fluid disperses the different spectra of colors as white light passes through that region of the crystal. They can also result from a simple crack in a crystal that varies in width and produces the same effect.

Fig. 172. *Several rainbows in Herkimer diamonds. Ace of Diamonds mine, Middleville, NY. 8 cm. Jay Walter collection.*

Smoky Crystals and Phantoms

Although Herkimer diamonds have a reputation for being colorless, they are also found in various shades of brown, commonly referred to as "smokies." Smoky quartz is not uncommon in the mineral world. On the contrary, it is actually very common. In Herkimer diamonds, however, it is not the norm. Most locations do not produce many smokies and this color variation is often proportional to crystal size with the larger crystals exhibiting darker smoky tints more frequently. Small crystals of a centimeter or less seldom show this smoky coloration. An exception to this rule are the crystals from Lowville, NY, and the Blair Hill farm and Grant quarry of Ontario, Canada that seem to have a higher frequency of smoky crystals even in smaller-sized crystals.

In metamorphic and igneous rock environments, it has been shown that smoky coloration in quartz is often the result of localized radiation that produces color centers where electrons are displaced at sites of Al substitution for Si in the structure. However, in Herkimer diamonds, smoky coloration is attributed to the inclusion of hydrocarbons that are present during the growth of the crystals. These microscopic hydrocarbons get trapped in the crystal's structure during its growth and produce a darker tint, or discoloration, within the normally colorless quartz. In calcite crystals, the hydrocarbon can lead to a diversity of colors ranging from yellow to orange to red, as well as brown.

Why the phenomenon is commonly more pronounced in larger crystals and the stems of scepters is likely due to the many phases of hydrocarbon involvement during the growth of the crystals in this region. Some of the results of this sporadic influx of hydrocarbon are: 1) scepter stems being highly included with later overgrowths being colorless, 2) smoky phantoms of standard shape in colorless crystals, 3) multiple smoky phantoms of standard shape in colorless crystals, 4) hydrocarbon phantoms in calcite and as coatings between different generations of calcite in locations such as Lowville and the Grant quarry, and 5) colorless quartz on drusy black crystals. Since larger crystals that have been growing for longer periods of time would be more likely to have been subjected to hydrocarbons mixing with quartz during their growth, they may more often be smoky. They are also thicker, making the same local concentrations of hydrocarbon inclusions more visible than in smaller crystals.

Phantoms are another feature common to the mineral quartz, yet uncommon in Herkimer diamonds. During the crystal's growth, material might coat all exposed surfaces, or in some cases, just selective faces on the crystal. This creates a frosted appearance. As clear quartz continues to be deposited on the crystal faces, these materials become trapped within the crystal yet remain clearly visible through the clear

Fig. 173. *A beautiful example of a Tessin-habit phantom in a water-clear quartz crystal. Brazil. 8.8 cm.*

Fig. 174. *Smoky phantom in a pair of crystals. Herkimer Diamond Mines Resort, Middleville, NY. 2 cm.*

Fig. 175. *Herkimer diamond with light smoky phantom. Ace of Diamonds mine. 5 cm.*

quartz that has grown after the impurity was deposited. The result is called a phantom, which is a visible shadow of the crystal's shape seen within the crystal on planes, often but not necessarily, parallel to the surface of the crystal. In Herkimer diamonds, they are almost always of a smoky color, again due to hydrocarbons trapped during crystal growth. Sometimes Herkimer diamonds will have phantoms created by other included materials such as sulfide particles.

Other Features Not Found in Herkimer Diamonds

There are other features found in quartz that appear to be absent in Herkimer diamonds. One characteristic found in alpine quartz locations around the world is the Faden. Fadens are lines found inside parallel growth quartz specimens that are attributed to tectonic movement that fractures the quartz with subsequent healing of the fracture. When repeated many times, a thin white line develops in otherwise clear quartz. Faden are typically found in quartz that formed in seams and veins. A second feature is called a Gwendal — a slow, gentle curve that can be found in what would have been parallel-growth quartz. Neither form is present in Herkimer diamonds probably because they formed in isolated cavities in a sedimentary rock host.

Curved Surfaces and Spheres

In 1981, Gait described curved surfaces on a crystal from the Hickory Hill locale as being pseudo-skeletal. Although perhaps better described as casts, these curved surfaces

Fig. 176. *Smoky, skeletal Herkimer diamond-style quartz with calcite. Lowville, Lewis County, NY. 6.5 cm.*

are correctly interpreted as coming from the formation of the quartz following patterns in pre-existing solidified hydrocarbons. In fact, these features are present at nearly all the Herkimer diamond occurrences in the traditional Herkimer diamond region and surrounding areas.

The original hydrocarbons were in liquid form allowing for flow features, fillings in fractures, and the formation of spherical voids due to degassing during solidification. In some cases, quartz later filled these voids and when not exposed to open space, followed the openings present in a cast-mold relationship. The resulting curved surfaces are most common on the base of crystals that formed on top of flows of anthraxolite. Quartz surfaces growing into open space would develop traditional faces that follow typical crystallographic parameters; thus, the creation of crystals with both curved and flat faces.

Small quartz spheres to two centimeters in diameter have been found at several Herkimer diamond locations. They normally are found within massive hydrocarbons and have slightly frosted surfaces. On occasion they show partial face development in the form of small circular flat surfaces corresponding to the locations where normal crystallographic faces should be — six surfaces matching the top terminal faces, six surfaces matching the prism faces, and six more surfaces matching the bottom termination faces. Other forms of circular surface development are relatively common and include microscopic divots and macroscopic concave and convex surfaces. These also seem to correspond to surfaces that were once in contact with hydrocarbon deposits.

The smallest of these features — spheres — are most likely due to small gas bubble fillings by quartz after hydrocarbon solidification. Those with partial face development would result from incomplete filling of the void during crystal growth.

Fig. 178. *Hydrocarbon (anthraxolite) embedded in areas of small surface divots on crystal surface. Collected 2002. Treasure Mountain mine, Fall Hill, Little Falls, NY. Field of view, 5 mm.*

Fig. 177. *Herkimer diamonds and calcites forming post depositionally around a hydrocarbon plate. Surfaces in contact with the hydrocarbon will follow any features present on the surface of the hydrocarbon. Paragenesis is hydrocarbon followed by quartz followed by calcite. Treasure Mountain mine, Fall Hill, Little Falls, NY. Collected 2002.*

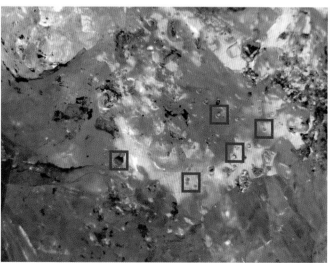

Fig. 179. *Gas bubbles in a large curved surface on bottom side of a Herkimer diamond. Treasure Mountain mine, Fall Hill, Little Falls, NY. Collected 2001. Field of view, 4 cm. The red squares indicate gas bubble impressions.*

Surface Growth Features

Some people take great interest in surface growth features on minerals. Generally speaking, Herkimer diamonds are relatively free of noticeable growth defects and present themselves with nice, clean surfaces relatively free of imperfections. On close inspection, however, growth deformations are relatively common on the faces of crystals found at all locations. Striations are occasionally present but are usually minimal and shallow. Striations are parallel grooves that, in the case of quartz and Herkimer diamonds, can be found on the prism faces of the crystal running perpendicular to the length of the prism (c-axis). Growth hillocks, small protuberances on crystal faces reflecting internal geometric symmetry in growth, can also be found but are uncommon as well.

Although quite common on quartz from other geological settings, secondary growth crystals on Herkimer diamonds are relatively rare. When present, they are usually regarded as a distraction from the otherwise normal perfection of these crystals and as such are not considered a desirable trait.

Fig. 182. *Herkimer diamond with secondary growth crystals and contact marks from secondary growth crystals. Crystal Grove mine, near Lassellsville, NY. 2.4 cm. Collected April 30, 2012.*

Fig. 180. *Surface growth features, giving the appearance of striations on prism face of Herkimer diamond. Lowville, NY.*

Fig. 181. *Unusual surface growth features found at the base of two different crystals. These are common on the prism faces of quartz and will point left or right depending on the handedness of the crystal. These are occasionally found on Herkimer diamonds but on the terminal faces pointing toward the prism faces. Treasure Mountain mine, Fall Hill, Little Falls, NY. Field of view of each image is approximately 5 millimeters. Collected 2002.*

Faceting

Sometimes Herkimer diamonds end up being cut into gemstones. This is generally done to damaged material that cannot be preserved as decent natural mineral specimens. The color of finished gemstones can vary from water clear to smoky. Material without distracting inclusions is preferred as with most forms of other gem rough. Some exceptional examples are shown below.

Faceted gemstones can, of course, be collected; however, their primary purpose is to be used in jewelry. Natural crystals are also mounted in jewelry or wire wrapped.

Fig. 183. *Hexagonal shaped flower cut weighing 210 carats. Herkimer Diamond Mines Resort. Stuart Strife collection.*

Fig. 187. *Portuguese round cut weighting 52.5 carats. Herkimer Diamond Mines Resort. Stuart Strife collection.*

Fig. 184. *Radiant cut weighing 56.5 carats. Herkimer Diamond Mines Resort. Stuart Strife collection.*

Fig. 185. *Trilliant cut weighing 35.5 carats. Herkimer Diamond Mines Resort. Stuart Strife collection.*

Fig. 186. *Modified radiant cut weighting 44.8 carats. Herkimer Diamond Mines Resort. Stuart Strife collection.*

Chapter 10:
Other Associated Minerals

There are associated minerals found with the high-quality quartz crystals of the Herkimer diamond region. Although normally secondary to the Herkimer diamonds, these other minerals add another level of interest for collectors and, in some cases, even enhance specimens to the point of their being the reason that collectors frequent one location over others.

Calcite

The most common accessory mineral within the dolostone is calcite, which forms crystals to 25 centimeters or more in diameter, depending on the locality. The problem with calcite is that it dissolves in even slightly acidic groundwater and is, therefore, often highly weathered. As a result, the calcite crystal surfaces are usually uneven and more or less dull in luster. Thus, the presence of calcite with the Herkimer diamonds can sometimes detract from the appearance of a specimen. This problem is often solved by removing the calcite entirely through immersion of the specimen in dilute hydrochloric acid. Some calcites, like those found at the

Benchmark quarry in St. Johnsville, are noteworthy and defy the norm. They can be quite gemmy, with exceptional luster, interesting form, and can contain inclusions of other minerals that are quite remarkable.

Calcite twins appear to occur at only one of the Herkimer or Herkimer-style quartz locations — Lowville.

Fig. 188. *Weathered calcites with phosphorescent calcite overgrowths on dolomite. Treasure Mountain mine, Little Falls, NY. 8.1 cm. Collected Summer 2002.*

Fig. 189. *Calcite with pyrite and dolomite. Benchmark quarry, St. Johnsville, NY. 10 cm.*

Fig. 190. *Calcite with marcasite on dolomite. Benchmark quarry, St. Johnsville, NY. 9 cm.*

Fig. 191. *Scepter calcites on dolomite. Lowville, NY. Field of view, 6 cm.*

Fig. 192. *Rhombohedral overgrowths on schalenohedral calcites of several colors with pyrite Grant quarry, Ontario, Canada. 4 cm.*

Fig. 193. *Calcite twin. Lowville, NY. Twinned crystal is 2.6 cm.*

Fig. 194. *Quartz on dolomite. Norwood quarry, near Norwood, NY. 8.4 cm.*

Dolomite

Dolomite in crystals to over a centimeter each are occasionally found in Herkimer diamond pockets and is plentiful at some localities. Most often it forms first in the order of crystallization (paragenesis) so that Herkimer diamonds are attached to dolomite crystals. Sometimes dolomite co-crystallizes with the quartz and, thereby, becomes an inclusion. Also, dolomite may form after the quartz, forming crystals on the faces of quartz. The range in quality is dependent on locality with some having dolomite crystals of brighter color and a higher luster, such as those from the Middleville quarry and those from the Barrett quarry near Norwood in St. Lawrence County. Most often, crystallized dolomite is found in saddle-shaped rhombohedral crystals of a gray to buff color and low luster.

Fig. 195. *Pyrite on dolomite. Lowville, NY. 8.4 cm.*

Fig. 196. *Herkimer diamonds on dolomite crystals. Benchmark quarry, St. Johnsville, NY. 7.3 cm.*

Fig. 197. *Parallel growth dolomite crystals on quartz with calcite. Lowville, NY. 6.1 cm.*

Dolomite sometimes fills some or all of a pocket's surfaces as a druse. When Herkimer diamonds or other accessory minerals form on these beds of dolomite crystals, the results can be very aesthetic.

Iron Sulfides

Pyrite is occasionally found as an accessory mineral with Herkimer diamonds. At several sites, it occurs in drapery and stalactitic forms deposited from the ceiling downward within pockets. In some cases, these forms are found in jumbled masses with other minerals in the pockets and may even show regrowth. More often, they can be seen as inclusions within the Herkimer diamonds, even sometimes extending out of the crystals. Distinct pyrite crystals have been observed but are uncommon and always small.

Marcasite is the other abundant iron sulfide in the Little Falls Dolostone. Although far from common, when present it can be quite aesthetic. It will usually appear as acicular needle-like inclusions within the Herkimer diamond

Fig. 198. *Herkimer diamonds on pyrite stalactites. Treasure Mountain mine, Little Falls, NY. 6.7 cm. Stuart Strife collection.*

Fig. 199. *Herkimer diamond with pyrite stalactites. Benchmark quarry, St. Johnsville, NY. 9.6 cm.*

Fig. 200. *Unusually formed pyrite crystals with step growth faces, quartz, calcite and dolomite. Lowville, NY. Crystals to 0.8 cm each.*

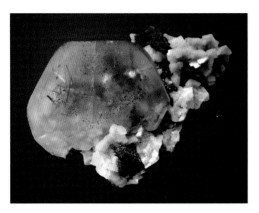

Fig. 201. *Marcasite inclusions in and on calcite crystal. Benchmark quarry, St. Johnsville, NY. 7 cm.*

Fig. 202. *Marcasite crystals on calcite. Benchmark quarry, St. Johnsville, NY. Field of view, 2.5 cm.*

Fig. 203. *Marcasite as inclusions. Benchmark quarry, St. Johnsville, NY. Field of view, 3 cm.*

crystals or other accessory minerals and sometimes extend out of those crystals as well. Marcasite crystals are brassy yellow in color and up to two or more centimeters in length.

Hydrocarbons

Hydrocarbon, a bitumen-like material, is present at all locations in a diversity of forms. The hydrocarbons found within the Herkimer diamond region and surrounding areas, usually referred to as anthraxolite, are commonly present in Herkimer diamond pockets, crystals, and the rock surrounding them. Several studies have been conducted on the material and the name anthraxolite has persisted with collectors, as a useful geological term.

Anthraxolite is usually a dark, lustrous, black color, unless it is weathered, and it can take many different forms. Flows of rich black anthraxolite are uncommon while stratified layers, beads, draperies, and even complete pocket fillings of the material are not uncommon. It is very common as inclusions in the quartz crystals and usually considered to be an impediment to forming the perfect, clear, euhedral specimens that are most desired by collectors.

Two strategies for the formation of anthraxolite have been suggested. Some believe that these hydrocarbon residues are a result of stromatolite decomposition and have thus been present in some form from original burial. Others think that fluid migration episodes have moved the hydrocarbons into the pockets. The fact that the hydrocarbons are found prior to, during, and after the formation of Herkimer diamonds results in a complex scenario. To have periods in which clear, un-included quartz has formed and other periods when heavy hydrocarbon inclusions dominated within the same pocket could imply multiple phases of fluid movement. This fits well with what is known of fluid migration and the movement of hydrocarbons such as oil. If the singular source was the

Fig. 204. *Massive hydrocarbon fragments (anthraxolite). Treasure Mountain mine, Little Falls, NY. Fragments to 3 cm each. Collected Summer 2002.*

Fig. 205. *Hydrocarbons embedded on druse. Herkimer Diamond Mines Resort, Middleville, NY. 3.7 cm.*

stromatolites' decomposition, there would be hydrocarbon available during the initial quartz growth but when depleted, it should not be available later on.

However, the actual formation of the quartz and anthraxolite might have been fairly complex. If the precursor material for the quartz was silica being held in solution by organic acid complexes and the precursor material for the anthraxolite would also have been a liquid, it is possible that the slow heating of the rock from deeper and deeper burial by overlying sediments might have had differential effects on the release of silica to form quartz crystals and the solidification of the anthraxolite. Moreover, these two processes could have interacted in complex ways.

Liquid "proto-anthraxolite" would have permeated cracks, porous rock, and vugs. With subsequent regional changes in temperature, the material would degas and solidify. Films, fillings, flows, and other forms of deposits would have undergone a reduction in volume and shrunk, and resulting cracks and fractures, mostly concoidal in form, would be preserved. At all these phases of change, the quartz can be shown to have been crystallizing and further, it can be shown that some areas had liquid hydrocarbons present after quartz crystals were fully formed.

Phase	Evidence/Examples
Pre-crystallization	Quartz forming in hydrocarbon fractures following openings
	Quartz spheres formed within gas bubbles
	Quartz forming on hydrocarbons creating casts
During crystallization	Hydrocarbon inclusions
Post-crystallization	Coatings on fully formed crystals
	Openings with hydrocarbon fully encasing crystals

Fig. 207. *Hydrocarbon flow partially overgrown by calcite. Treasure Mt. mine, Fall Hill, Little Falls, NY. Field of view, approximately 9 cm. Collected Summer 2002. The red squares indicate imprints of gas bubbles.*

Fig. 206. *Hydrocarbon flow. Treasure Mountain mine, Fall Hill, Little Falls, NY. 5.5 cm. Collected Summer 2002. Steve Chamberlain collection, #15457 and photograph.*

Fig. 208. *Hydrocarbon plate. Treasure Mountain mine, Fall Hill, Little Falls, NY. 3 cm. Collected Summer 2002.*

Fig. 209. *Hydrocarbon covering massive dolomite plate with dolomite crystals. Treasure Mountain mine, Fall Hill, Little Falls, NY. 10 cm. Collected Summer 2001.*

Fig. 210. *Close-up of hydrocarbon plate (Fig. 218) with dolomite. Treasure Mountain mine, Fall Hill, Little Falls, NY. Field of view, 3 cm. Collected Summer 2001.*

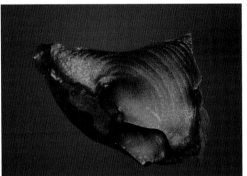

Fig. 211. *Hydrocarbon showing spherical shrinkage lines. Treasure Mountain mine, Fall Hill, Little Falls, NY. Left: 2 cm. Right: 2.5 cm. Collected Summer 2002.*

Fig. 212. *Hydrocarbon spherical void resulting from hydrocarbon shrinkage. Treasure Mountain mine, Fall Hill, Little Falls, NY. 2 cm. Collected Summer 2002.*

Fig. 213. *Hydrocarbon casts after dolomite. Treasure Mt. mine, Fall Hill, Little Falls, NY. 1.8 cm. Collected Summer 2002.*

Hydrocarbon may not be considered a mineral, but it is an important geological material shown to have been present before, during, and after formation of almost all of the mineral species found in the pockets in the region. Further, modern studies seem to indicate that the presence of the hydrocarbon may be intimately related to the formation of Herkimer diamonds so its importance cannot be overstated, even if specific relationships between the formation of quartz and anthraxolite have not yet been worked out.

Others

There are other documented cases of minerals such as gypsum, sphalerite, and celestine being found with Herkimer diamonds, but these examples are very rare. More common are the reports of odd minerals and other materials having been found within Herkimer diamonds as inclusions. Rumors of gold or plant fossils as inclusions abound. Those that have been scientifically investigated have been debunked but the rumors persist.

Chapter 11:
Final Thoughts

As comprehensive as the preceding chapters are, much more data and ideas could have been included if space permitted. Searches on the internet reveal vast amounts of further information; however, much of it is unedited and unconfirmed, so the reader must be somewhat skeptical. For some topics, such as what kind of lighting is available for displaying Herkimer diamonds, a search can reveal a whole day's reading of useful information.

The very large number of people who have collected Herkimer diamonds has given rise to a significant body of field-collecting stories — some published, some available on the internet, and many just part of the rich oral tradition of the mineral culture. Anecdotal experiences from searching for and finding Herkimer diamonds could fill another volume.

Interest in claimed metaphysical properties of minerals, especially quartz, remains high in American society. Herkimer diamonds are center stage in the metaphysical movement.

Since scientific studies of Herkimer diamonds and similar forms of quartz are ongoing, the amount of research data not covered here will continue to grow. Eventually, a more complete picture of the formation of Herkimer diamonds will emerge. For now, this book provides a window into the universe of knowledge about Herkimer diamonds tailored to those who wish to field-collect, assemble collections of, and appreciate these magnificent New York State crystals.

Appendices

Appendix A: Road Maps

Ace of Diamonds and Herkimer Diamond Mines and Resort

Both of these quarries are found near the southern edge of the village of Middleville on Route 28 and are well marked.

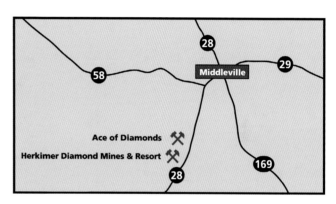

Fig. 214. *Map showing locations of Ace of Diamonds and Herkimer Diamond Mines and Resort south of Middleville, NY.*

Crystal Grove

The Crystal Grove mine is west of County Route 114 about 500 yards south of its intersection with Route 29. This intersection is about 1.5 miles west of the village of Lassellsville and 3 miles east of the village of Oppenheim. The mine diggings, store, and campgrounds are all well marked.

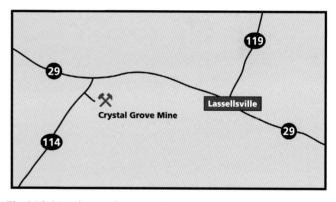

Fig. 215. *Map showing location of Crystal Grove west of Lassellsville, NY.*

Diamond Acres and Hickory Hill

Diamond Acres is south of County Highway 34 (Stone Arabia Road), just west of its intersection with Barker Road about 5 miles west of the village of Fonda. The drive is several hundred feet from the intersection and is not conspicuous.

The Hickory Hill Diamond Diggings are approximately 4.5 miles west of the village of Fonda on County Highway 33, just west of its intersection with Barker Road. The turn south into the site is difficult to see.

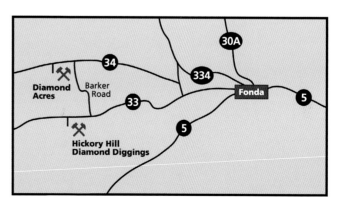

Fig. 216. *Map showing locations of Diamond Acres and Hickory Hill west of Fonda, NY.*

Appendix B: GPS Coordinates for Herkimer Diamond Locations

Ace of Diamonds mine	43° 07' 56" N (Lat.), 74° 58' 27" W (Long.)
Herkimer Diamond Mines and Resort	43° 07' 49" N (Lat.), 74° 58' 32" W (Long.)
Crystal Grove	43° 03' 05" N (Lat.), 74° 38' 02" W (Long.)
Diamond Acres	42° 57' 46" N (Lat.), 74° 28' 32" W (Long.)
Hickory Hill	42° 56' 33" N (Lat.), 74° 28' 13" W (Long.)
Hanson quarry, Middleville	43° 09' 11" N (Lat.), 74° 58' 37" W (Long.)
Treasure Mountain mine	43° 02' 09" N (Lat.), 74° 50' 50" W (Long.)
Benchmark quarry	43° 00' 35" N (Lat.), 74° 42' 04" W (Long.)
Gailor quarry	43° 05' 38" N (Lat.), 73° 46' 39" W (Long.)
Diamond Point/ Island	43° 27' 32" N (Lat.), 73° 40' 35" W (Long.)
Norwood quarry	44° 46' 51" N (Lat.), 74° 59' 32" W (Long.)
Ogdensburg quarry	44° 40' 57" N (Lat.), 75° 30' 27" W (Long.)
Lowville	Undisclosed
Grant quarry	45° 16' 38" N (Lat.), 75° 49' 01" W (Long.)
Blair Hill farm	45° 13' 30" N (Lat.), 75° 30' 59" W (Long.)

Appendix C: Crystal Drawings of Herkimer Diamonds (from Goldschmidt, 1986/1922)

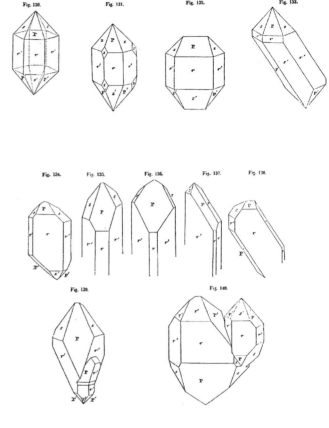

Fig. 217. *Crystal drawings of Herkimer diamonds from Goldschmidt (1986/1922), Figs. 130-140.*

Appendix D: The Anatomy of Scepters

The Scepter

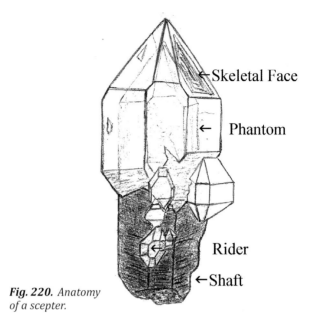

Fig. 220. *Anatomy of a scepter.*

The Dumbbell Scepter

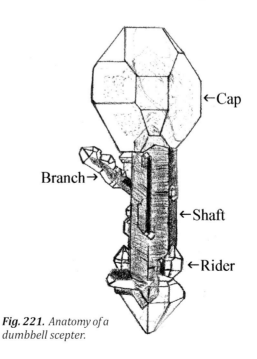

Fig. 221. *Anatomy of a dumbbell scepter.*

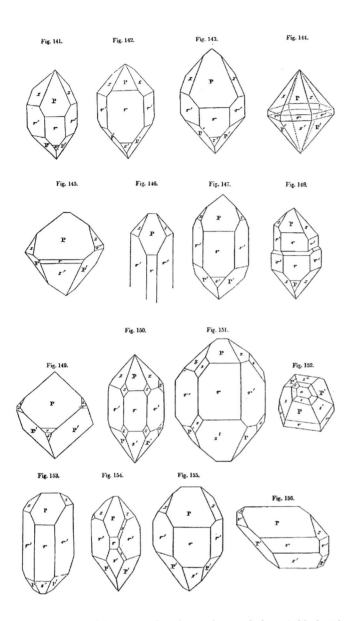

Fig. 218. *Crystal drawings of Herkimer diamonds from Goldschmidt (1986/1922), Figs. 141-156.*

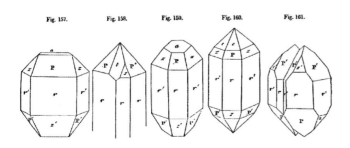

Fig. 219. *Crystal drawings of Herkimer diamonds from Goldschmidt (1986/1922), Figs. 157-161.*

The Internal Shaft Scepter

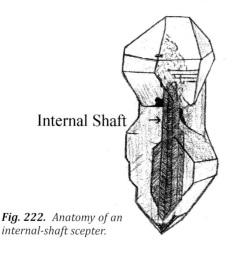

Internal Shaft

Fig. 222. Anatomy of an internal-shaft scepter.

The Complex Scepter Cluster

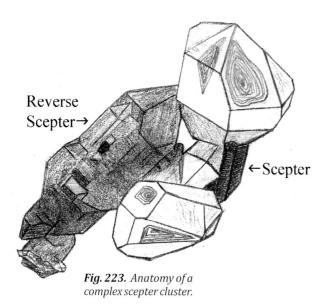

Reverse Scepter→

←Scepter

Fig. 223. Anatomy of a complex scepter cluster.

Appendix E: Analysis of Some Hydrocarbons (Anthraxolite)

	1	2	3	4
C	90.42	90.50	90.25	93.5
H	3.94	3.97	4.16	2.81
O	3.42	1.93	3.69	2.72
N	1.30	1.36	0.52	0.97
S	0.57	0.52	0.66	—
Ash	0.35	1.72	0.72	—
	100.00	100.00	100.00	100.00

1) Anthraxolite, uppermost Little Falls dolostone, Flat Creek near Sprakers, NY

2) Anthraxolite, Little Falls dolostone, abandoned quarry near Salisbury, NY

3) Anthraxolite, Black River limestone near Kingston, Ontario, Canada

4) Anthracite Coal, average analyses of 16 samples, multiple locations

(Modified from Dunn and Fisher, 1954)

Appendix F: Properties of Hydrocarbons (Anthraxolite)

Physical Properties:

Color	black, rarely brown, yellow
Hardness	2.0 (crusts at Little Falls)
Luster	vitreous to dull
Fracture	conchoidal
Streak	black
Specific Gravity	1.32–1.41
Fusibility	infusible
Crystal Form	amorphous
Structure	botryoidal, massive, crustiform

Chemical Properties:

1) Insoluble in carbon tetrachloride, benzene, acetone, toluene, ethyl ether, methyl alcohol, diethylene glycol, petroleum ether, dioxane, ethanolamine, and carbon disulfide

2) Decomposed by concentrated sulfuric acid

3) No lines in spectroscopic analysis of anthraxolite from Flat Creek that could not be explained as impurities in the carbon electrodes

4) Mostly unaltered in a heated closed tube test

5) "Petroliferous" odor emitted when material from the spoil bank at Little Falls is crushed

6) Two types of hydrocarbon in a Middleville quartz crystal: a yellow type melting at 70-80° C, and a brown type melting at 200-220° C

(Modified from Dunn and Fisher, 1954)

Glossary

Adit — A horizontal mine opening or tunnel into rock.

Anthraxolite — A coal-like, hydrocarbon-based material which forms in Herkimer diamond pockets and sometimes as inclusions within the crystals.

Arch — A semicircular group of interconnected Herkimer diamond crystals.

Barbell (syn. Dumbbell) — Scepter having terminations (caps) at both ends taking on the shape of a dumbbell or barbell used in weight lifting.

Black Crystals — Crystals densely filled by black anthraxolite.

Black Druse Pocket — A pocket containing colorless crystals on a lining of tiny, dark black, anthraxolite-included crystals. Very rare.

Blowout — 1) A circular chip out of a crystal face that is often the result of a frozen enhydro. 2) When a spring-steel stack explodes out of the crack in which it was secured.

Blue Rock — A hard, dense, silica-rich dolostone. Often used to describe the hardest rock in the quarry.

Cabinet Specimens — Specimens which are approximately 10 to 15 centimeters in size.

Cap — The termination of a scepter (see Appendix D).

Capeing — The cutting of a groove into the rock to serve as a crack in which to set spring-steel wedges.

Capeing Tool — A specialized chisel used to cut slots for spring-steel wedges.

Carbon Pocket — A pocket with a very high concentration of, or totally filled by, loose hydrocarbons (anthraxolite).

Cavity — An opening in a rock where crystals can form. A synonym for vug or pocket.

Charcoal Pocket — A pocket containing large amounts of hydrocarbon material (anthraxolite).

Chatter Marks — A series of tiny dings across a crystal's edge or surface.

Claim — 1) A location where mining is reserved for those who hold, and usually pay for, the right to collect there. 2) A legal agreement made with a governing body to mine minerals from a location.

Claim Jumper — An individual who collects on someone else's claim without permission.

Counterfeit — The description given to a mineral specimen that is in some manner constructed. Examples include, but are not limited to, crystals glued together in unnatural forms, crystals attached to pieces of matrix where they did not form or in positions that are unnatural. Syn. Fake

Country Rock — The bedrock native to a region or area.

Clusters — Groups of intergrown crystals.

Crack Hammer — A rock hammer designed with an enlarged head. Similar to a sledgehammer but smaller.

Crosses — Crystals in the shape of a crucifix.

Crystal — The geometric shape formed when a mineral can grow in an unrestricted space. The faces and angles reflect the orderly arrangement of atoms inside.

Ding — A tiny chip on a crystal's surface.

Dome — A raised center in a pocket.

Doughnut — A circular cluster of Herkimer diamonds with an opening in its center.

Draperies — Sheets of pyrite like draped cloth found in Herkimer diamond pockets and within crystals with calcite. Similar to cave formations of calcite that go by the same name.

Druse — Coating of tiny quartz crystals on a rock matrix; part of the lining of a pocket.

Dry Pocket — A vug or opening within the rock that contains no crystals.

Dumbbell (syn. Barbell) — Scepter having terminations (caps) at both ends taking on the shape of a dumbbell or barbell used in weight lifting.

Dump — The piles of waste rock or unprocessed ore left at a mine site.

Elestial — A crystal exhibiting extensive skeletal growth.

Enhydo — A liquid inclusion.

Epitaxis — An overgrowth of crystals of one species of mineral which follows the crystal geometry of an underlying, and different, species of mineral.

Euhedral — A crystal with well-defined flat faces and no or minor attachment points.

Face — A plane which defines the external shape of a crystal.

Fake — The description given to a mineral specimen that is some manner constructed. Examples include but are not limited to crystals glued together in unnatural forms, crystals attached to matrices they did not form on or in positions that are unnatural. Syn. Counterfeit.

Fixed — When a crystal is attached to the wall rock.

Flat — The flat cardboard boxes used to transport some sodas and alcoholic beverages or those especially made for the mineral specimen industry.

Floater — A crystal with faces on all sides and no evident points of attachment. Also, applies to groups of crystals with no points of attachment.

Fracture — A crack in the rock or in a crystal.

Habit — The characteristic appearance of a crystal as determined by its predominant crystal forms.

Inclusions — Solids, liquids and or gases trapped within a crystal of a mineral.

In situ — In place. Herkimer diamonds found on the walls of the pocket can be referred to as being *in situ*.

Jewelry Pocket — A pocket with a mostly filled by loose anthraxolite and small, jewelry-grade (high-quality/transparent) Herkimer diamonds.

Junkers — The largest pockets that contain larger, more fractured crystals. (Middleville specific).

Leaverites — Specimens that are so poor in quality you should "leave er right" where you found them.

Ledge or **Ledge Rock** — The layer of rock found below the table layer that contains sub-pockets (Middleville specific).

Lift — An area of rock that is being broken free from its bedding plane usually by using steel.

Massive — A mineral with no visible crystal form.

Matrix — The rock on which, or within which, crystals form.

Metamorphism — The process of change in a rock resulting from elevated temperature and pressure.

Micromounts — Tiny crystals best viewed under a microscope.

Miner — A term used regionally, and in this book, to describe active field collectors of mineral specimens as well as its traditional usage to describe those employed within more formal mineral extraction operations at mines and quarries.

Miniatures — Specimens in the 3 to 6 cm range.

Natural — A cluster of Herkimer diamonds that does not require reassembly.

Overburden — The soil and shatter rock found above the zone in which the crystals are located.

Paragenesis — The order of formation of minerals.

Phantom — A shadow visible within a transparent or translucent crystal created by microscopic inclusions on preexisting crystal faces.

Pocket — An opening in a rock where crystals can form. Syn. vug or cavity.

Prism — The faces that make up the sides or the body of the crystal.

Reassembly — When clusters of loose Herkimer diamonds are reattached to one another in their original orientations using adhesives.

Reattachment — A form of repair where a crystal dislodged from its matrix is glued back into its original position.

Refractive Index — The ratio of the speed of light in a vacuum to its speed in a solid or liquid. Indicates how much light will be bent at the interface of two different media.

Repair — When a broken crystal is glued back together or a crystal is reattached to its original matrix.

Restoration — When a crystal or mineral specimen is restored to the condition it was believed to have had in nature.

Reverse Scepter — A crystal growth form that tapers at the termination.

Rider — An attachment crystal on a scepter or larger Herkimer diamond (see Appendix D).

Rods — Thin stalactites of pyrite occasionally found within quartz or calcite from the Herkimer diamond locations.

Scepter — A crystal growth form with an enlargement (cap) at its termination. The form resembles the scepters of royalty.

Seam — A fracture in the rock sometimes filled by minerals that were deposited at a later time.

Shaft — The stem section of a scepter crystal (see Appendix D) or vertical entrance into a mine.

Share Diggers — The miners who are hired by the mine to collect crystals and are paid with a share of what they find.

Shatter Rock — The layers of rock found above the table layer in the rock strata (Fonda/ Diamond Acres specific).

Silver Picking — Buying specimens instead of mining them yourself.

Singing — The sound a spring-steel wedge makes when well secured in the dolostone and being pounded on by a sledgehammer.

Skeletal — Having crystal faces which do not appear fully developed. They look cut up in appearance due to the crystal's edges developing faster than its faces.

Smokies — Herkimer diamonds with a brownish, smoke-like tint due to microscopic hydrocarbon inclusions.

Specific Gravity — The number that expresses the ratio between the weight of the substance and the weight of an equal volume of water (the relative density of the material).

Stack — Several sections of spring-steel wedges set in a crack and used together to break dolostone.

Steel — Short for **Spring-Steel Wedges** that are made from the leaf springs found in the suspension systems of automobiles.

Striations — Shallow, parallel grooves on the crystal's surfaces.

Stromatolites (Stroms) — Ancient marine cyanobacteria which formed mats or domes that are sometimes preserved as fossils within dolostone. Stromatolites are believed to be the source of the pockets found in the Herkimer diamond region.

Subhedral — A crystal having faces on some, but not all, of its surfaces.

Sub-pockets — Smaller pockets found below the table layer that often contain smaller, but higher quality, Herkimer diamonds.

Suite — A grouping of all the mineral species from a given location.

Tabular (Tabby) — A crystal exhibiting a flattened appearance.

Table or **Table Layer** — The layer of sedimentary rock in which the primary pockets form. Usually the hardest layer in the rock strata.

Tailings — The waste rock and soil left behind by previous collectors at a dig site.

Talking — The sound of dolostone cracking while being mined.

Temper — The hardening of steel desired for spring-steel wedges or chisels.

Termination — The end of the crystals where the prism faces grow together.

Thermal Shock — Effects of dramatic changes in temperature on crystals. It can cause internal haziness or breaking.

Three-Phase Inclusions — Inclusions composed of liquid and gas bubble(s) and solids such as hydrocarbons (anthraxolite) or rarely calcite or dolomite crystals.

Thumbnail Specimens — Specimens which are less than 2.54 cm in size. For competition these specimens must be able to fit into the volume of a one inch cube.

Twins — Pairs of crystals that grew in specific geometric orientation according to one of the twin laws for that mineral.

Two-Phase Inclusions — Inclusions composed of liquid and gas bubble(s).

Undermining — The breaking away of the table layer without the removal of the overburden. The process, when extensive, can result in dangerous overhanging ledges of rock.

Vug — An opening in a rock where crystals can form. Syn. pocket or cavity.

Wall — The working face of vertical rock in a quarry or mine.

Wall Rock — The rock making up the inside of the pocket or vug.

References

BACK, M. E. and MANDARINO, J. A. (2008) *Fleischer's Glossary of Mineral Species, 2004*. The Mineralogical Record, Inc. Tucson.

BANCROFT, P. (1973) *The World's Finest Minerals and Crystals*. The Viking Press, New York, p. 176.

BECK, L. C. (1842) *Mineralogy of New-York*. Albany, New York: W. and A. White and J. Visscher.

BENNETT, P. C. (1991) Quartz dissolution in organic-rich aqueous systems. *Geochimica et Cosmochimica Acta* 55:1781-1797.

BENNETT, P. and SIEGEL, D. I. (1987) Increased solubility of quartz in water due to complexing by organic compounds. *Nature* 326:684-686.

BOROFSKY, R. L., WHITMORE, R. and CHAMBERLAIN, S.C. (2000) Scepter Quartz Crystals from the Treasure Mountain Diamond Mine, *Rocks & Minerals* 75:231-237.

BROWN, C. E. (1983) *Mineralization, mining, and mineral resources in the Beaver Creek area of the Grenville Lowlands in St. Lawrence County, New York*. U.S. Geological Survey professional paper 1279.

CHAMBERLAIN, S. C. (1988) On the origin of "Herkimer diamonds." *Rocks & Minerals* 63:454.

CHAMBERLAIN, S. C., and HLADYSZ, W. J. (1997) Black-stemmed 'Herkimer Diamond' scepters from Fall Hill, Little Falls, Herkimer County, New York. *Rocks & Minerals* 72:121.

CUSHING, H. P. (1905) Geology of the Vicinity of Little Falls, Herkimer County. New York State Museum Bulletin #77.

DANA, E. S. (1877) *A Textbook of Mineralogy*. John Wiley and Sons, New York.

DANA, E. S. (1895) *Minerals and How to Study Them*. John Wiley and Sons, New York.

DIETRICH, R. V. (1956) *Economic Geology*, Society of Economic Geologists 51:649-664.

DOSSERT, W. P., and CHAMBERLAIN, S. C. (1991) "Herkimer diamond-like" quartz mineralization and calcite twins in the Marcellus shale near Syracuse, Onondaga County, New York. *Rocks & Minerals* 66:41-42.

DUNN, J. and FISHER, D. (1954) Occurrence, properties and paragenesis of anthraxolite in the Mohawk Valley. *American Journal of Science*, 252:489-501.

FAST, J. B. (2008) A 2007 collecting venture in Herkimer County, New York: Thirty-eight quartz pockets. *Rocks & Minerals* 83:196-200.

FRANCIS, B. (1953) The Largest Herkimer Diamond. *Rocks & Minerals* 1:26-27.

FRIEDMAN, G. M. (1987) Vertical movements of the crust: case histories from the northern Appalachian basin. *Geology*, 15, p. 1130-1133.

GAIT, R. I. (1981) Pseudo-skeletal quartz crystals from Hickory Hill, Montgomery County, near Fonda, New York. *Rocks & Minerals* 56:153-155.

GAULT, H. R. (1949) The frequency of twin types in quartz crystals. *American Mineralogist* 34:142-162.

HADLEY, J. (1832) *New-York Medical and Physical Journal*, p. 132.

HLADYSZ, W. J, HLADYSZ, V. J., and CHAMBERLAIN, S. C. (1997) Black-stemmed 'Herkimer Diamond" scepters from the Eastern Rock Products Quarry, St. Johnsville, Montgomery County, New York. *Rocks & Minerals* 72:125.

HURLBUT, C. S., and KLEIN, C. (1977) *Manual of Mineralogy*, 19th ed., John Wiley & Sons, New York, p. 3.

ISACHSEN, Y. W., and FISHER, D. W. (1970) Geologic Map of New York, Adirondack Sheet. University of the State of New York, The State Education Department, Geological Survey.

JENSEN, D. E. (1978) *Minerals of New York State*. W. F. Humphrey Press, Geneva, New York.

JONES, B. (1976) Herkimer diamonds. *Rock & Gem* 6:20-26, 76-77.

JONES, B. (2009) Crystals so gemmy they're called "diamonds." *Rock & Gem* 39:12-16.

KAPELEWSKI, J. A. (2009) Tails from decades of Herkimer diamond digging. *Rock & Gem* 39:22-25.

KEITH, M. T. and TUTTLE, O. F. (1952) Significance of variation of quartz in the high-low inversion of quartz. *American Journal of Science*, Bowen Volume, p. 244.

KLEIN, C. (2002) *The Manual of Mineral Science, 22nd edition*. John Wiley & Sons, Inc., p. 6.

KNOPF, E.B. (1927) The Little Falls Formation. *American Journal of Science Science*, 5th ser., 14:429-458.

KUNZ, G. W. (1892) *Precious stones of North America*. Dover Publications, New York.

LA BUZ, A. L. (1969) The "Herkimer diamond" grounds. *Rocks & Minerals* 44:243-250.

MCELWEE, M. A. (1999) Minerals of the Lockport Formation. *Rocks & Minerals* 74:185-86.

MEDICI, J. (1996) The Smithsonian gets a Herkimer diamond pocket. *Rocks & Minerals* 71:397-401.

MITCHELL, J. R. (1982) Field trip: Herkimer diamonds: six locations where you can dig New York's quartz "diamonds." *Rock & Gem* 12:62-67.

MOORE, B. S., 1989. *Herkimer Diamonds, A Complete Guide for the Prospector and Collector*. Phoenix Printing Company, p. 4-5.

MUSKATT, H.S. and TOLLERTON, V.P., Jr., 1992, The Little Falls Dolostone (Late Cambrian); stratigraphy and mineralogy. In: April, R.H., ed., Field trip guidebook: New York State Geological Association Guidebook, no. 64, 64th Annual Meeting, Hamilton, NY, September 18-20, 1992, p. 200-215.

NEW YORK STATE MUS. BULLETIN (1905) no. 77, map.

NEW YORK STATE MUS. BULLETIN (1910) no. 140.

NEWLAND, D. H. (1937) Herkimer County, New York Quartz Crystals. *Rocks & Minerals* 12:36-37.

ROBINSON, G. W. (1994) *Minerals: An Illustrated Exploration of the Dynamic World of Minerals and Their Properties.* Simon & Schuster, New York.

ROBINSON, G. W., Dix, G. R., Richards, R. P., and Picard, M. (2011) Minerals of the Beekmantown group, Southeastern Ontario, Southern Quebec, and Northwestern New York. *Rocks & Minerals,* 86:546-560.

ROEDDER, E. (1979) Fluid inclusion evidence on the environments of sedimentary diagenesis, a review. *The Society of Economic Paleontologists and Mineralogists,* Special Publication, 26, p. 89-107.

ROWLEY, E. B. (1951) Crystal Collecting at Saratoga Springs, N.Y., *Rocks & Minerals,* 26:528-532.

SCHNEIDER, J. F. (1975) Recent tidal deposits, Abu Dhabi, UAE, Arabian Gulf. In: *Tidal Deposits,* Ginsburg, R. N. (ed.) Springer-Verlag, Berlin, p. 209-214.

SINKANKAS, J. (1964) *Mineralogy for Amateurs.* D. Van Nostrand Company, Inc., Princeton, NJ, p. 513.

SNOW, D. (1996) *The Iroquois.* John Wiley & Sons, New York, 271p.

TERVO, W. A. (1967) The *Rock Hound's Guide to New York State Minerals, Fossils and Artifacts.* Exposition Press, Inc., New York, New York, p. 38-39.

TUTTLE, D. L., (1973) Inclusions in "Herkimer diamonds," Herkimer County, New York quartz crystals. *Lapidary Journal,* 27:966-976.

ULRICH, E. O. and CUSHING, H.P., (1910) Beekmantown Formation. *Geological Society of America Bulletin* 21:780-781.

ULRICH, W. (1989) The quartz crystals of Herkimer County and its environs. *Rocks & Minerals,* 64:108-122.

VANDERBILT, H. L. (1985) Herkimer diamonds. *Lapidary Journal* 39:45-47.

VAN DIVER, B. B. (1976) *Rocks and Routes of the North Country, New York.* W. F. Humphrey Press, Geneva, New York, p. 153-154.

VAN DIVER, B. B. (1980) *Field Guide to Upstate New York.* W. F. Kendall/ Hall Publishing Company, Dubuque, Iowa.

WALKER, D. and WALKER, C. (1990) Herkimer "diamonds." *Lapidary Journal* 44:71-72.

WALLER, R. (1980) A rust removal method for mineral specimens. *Mineralogical Record,* 11:109.

WALTER, M. R. (2002) Finding Scepters at Treasure Mountain, Little Falls, New York. *Rock & Gem,* February issue, 32:64-69.

WALTER, M. R., (2003) How to make and use spring steel wedges. *Rock & Gem,* June issue, 33:60-61.

WALTER, M. R. (2004) Diamond Acres. *Rock and Gem* 34:64-66, 68-69.

WALTER, M. R. (2007) *Field Collecting Minerals in the Empire State: Stories of Modern Day North Country Miners.* Privately Published.

WILMARTH, M.G., (1927) Geologic Names Committee remarks on the Little Falls and Theresa formations, ca. 1905-1927]. In: Wilmarth, M.G., 1938, Lexicon of geologic names of the United States (including Alaska): *U.S. Geological Survey Bulletin,* 896, pt. 1, p. 1195-1196.